**High Frequency Conducted
Emission in AC Motor Drives
Fed by Frequency Converters**

High Frequency Conducted Emission in AC Motor Drives Fed by Frequency Converters

Sources and Propagation Paths

Jaroslaw Luszcz
Gdansk University of Technology, Poland

WILEY

Published by John Wiley & Sons, Inc., Hoboken, New Jersey.
Published simultaneously in Canada.

For general information on our other products and services or for technical support, please contact our Customer Care Department within the United States at (800) 762-2974, outside the United States at (317) 572-3993 or fax (317) 572-4002.

Wiley also publishes its books in a variety of electronic formats. Some content that appears in print may not be available in electronic formats. For more information about Wiley products, visit our web site at www.wiley.com.

Library of Congress Cataloging-in-Publication Data is available.

ISBN: 978-1-119-38839-5

Printed in the United States of America.

V071650_061118

To those who have not been given the chance to get the education they wanted.

When people of similar frequencies come together,
output is not a simple sum of individual work, but exponential.
In science we term this phenomenon as resonance.
Output at this stage is beyond any logical limit.

Ravindra Shukla

Contents in Brief

Contents

List of Figures

List of Tables

Preface

The understanding of generation and propagation of conducted emission in power electronic converters is on one hand already well recognized and characterized by using fundamental laws of physics. On the other hand, it is still a challenging task to resolve many of interference issues faced in contemporary real system, which from one year to another become more and more complex electromagnetically, because they tend to simultaneously include highly emissive high speed devices of significant power and exact control system which are electromagnetically sensitive. Solving interference problems in such environment requires not only knowledge of the basics of electromagnetism but also the strong skills needed for effective identification and selection of the most critical interference issues occurring in particular application.

The concepts presented in this book are focused mainly on finding and formulating the issues most likely to occur related to generation and propagation of conducted emission in AC motor drives fed by frequency converter, rather than proposing specific solutions for dealing with those problems. This book is intended for scholars and a wide range of professionals who are involved in the stages of development, design, and application of adjustable speed drives in accordance with the ever increasing EMC requirements. This book is the outcome of a long process, lasting over many years, of completing my education, conducting research, and attaining the knowledge and professional practical experience, which led me through this discipline and inspired me to discover some of the hidden mysteries of the science and art of EMC.

I would like to express my deepest gratitude to the many people who help me through this process. Without their encouragement, support, guidance, inspiring questions, comments, and the editing assistance this book would have never come into being.

Firstly, I would like to recall the memories of my early tutors who let me experience my first own electrical experiments, which boost me to this profession.

Secondly, I am especially grateful to all individuals who I met in my professional academic life, superiors and colleagues, for their mentorship, cooperation, assistance, helpful remarks, valuable advices and suggestions. Without them my EMC experience would not have been as it is. Equally so, I thank my friends, who kept on asking me year in year out, "When you finish?" only to receive the same response: "Soon.", and remained remarkably tolerant.

Thirdly, my appreciation also goes to the many students and engineers whom I have educated and cooperated in EMC area throughout the years, for their inquiring and thought-provoking questions, and excellently pointing out many atypical and realistic aspects of considered issues.

Finally, I would like to extend my acknowledgment to the technical reviewers and the editorial board of this book, for their involvement, efforts, and personal time spent for improving the book. Their expertise and experience have greatly enhanced the quality and clarity of this book.

Gdansk, Poland   Jaroslaw Luszcz

Symbols

C	capacitance in general
f	frequency in general
$f_{r,C}$	self-resonant frequency of capacitor
$f_{r,L}$	self-resonant frequency of inductor
$f_{Rp,L}$	the main parallel resonance of inductor
$f_{Rp,C}$	the main parallel resonance of capacitor
$f_{Rs,L}$	the main serial resonance of inductor
$f_{Rs,C}$	the main serial resonance of capacitor
$f_{T,L}$	transition frequency of inductor
$f_{C,T}$	transition frequency of cable
$f_{Rp,M}$	the main parallel resonance of AC motor winding
$f_{Rs,M}$	the main serial resonance of AC motor winding
f_1	fundamental frequency of periodic waveform
f_{PWM}	carrier frequency of PWM
I_{CM}	common mode current
I_{DM}	differential mode current
$I_{CM,In}$	input common mode current
$I_{CM,Out}$	output common mode current
$I_{CM,M}$	motor common mode current
$I_{CM,FC}$	frequency converter common mode current
$I_{CM,MC}$	common mode current of motor cable
$I_{CM,ShM}$	common mode current returning back from motor by feeding-cable shield
$I_{CM,Gnd}$	common mode current returning back from motor by ground connections, other than cable shield
$I_{CM,ShFC}$	common mode current returning back to FC by motor feeding cable shield
L	inductance in general
R	resistance in general
t_r	rise time

$U1, U2$	AC motor windings terminals – phase a
$V1, V2$	AC motor windings terminals – phase b
$W1, W2$	AC motor windings terminals – phase c
V_{CM}	common mode voltage
V_{DM}	differential node voltage
Z	impedance in general
Z_C	characteristic impedance
$Z_{CM,PG}$	common mode impedance of power grid, line-to-ground
$Z_{Gnd,FC}$	grounding impedance of frequency converter
$Z_{Gnd,M}$	grounding impedance of AC motor
$Z_{DM,PG}$	differential mode impedance of power grid, line-to-line
$Z_{Sh,MC}$	impedance of motor cable shield
V_{L-N}	phase voltage of power grid
V_{L-L}	line voltage of power grid

Greek Symbols

dv/dt	rate of change of voltage
ω	angular frequency in general
v	signal propagation velocity
μ_0	permeability of free space constant ($\mu_0 \approx 1.257 \times 10^{-6}$H/m)
μ_r	relative permeability
ε_0	permitivity of free space ($\varepsilon_0 \approx 8.854 \times 10^{-12}$F/m)
ε_r	relative dielectric constant
c_0	speed of light in free space ($c \approx 3 \times 10^8$m/s)
k_p	propagation factor
R_0	equivalent resistance per unit length
C_0	equivalent capacitance per unit length
L_0	equivalent inductance per unit length
λ	wavelength, the distance along the direction of propagation of a periodic wave between two successive points where, at a given time, the phase is the same

Acronyms

AC	alternating current
ASD	adjustable speed drive
CENELEC	European Committee for Electrotechnical Standardization
CISPR	International Special Committee on Radio Interference
CM	common mode
DC	direct current
DM	differential mode
EM	electromagnetic
EMC	electromagnetic compatibility
EMD	electromagnetic disturbance
EMI	electromagnetic interference
FC	frequency converter
Gnd	"Ground" as a reference electric zero potential with respect to the earth
HF	high frequency
IEC	International Electrotechnical Commission
IEEE	Institute of Electrical and Electronic Engineers
IGBT	insulated gate bipolar transistor
LISN	line impedance stabilization network
LV	low voltage
LF	low frequency
MV	medium voltage
M	AC motor
MC	AC motor feeding cable
PE	protective earth connection, earthing terminal in electrical devices
PG	power grid

PQ	power quality
PWM	pulse width modulation
RF	radio frequency
RFI	radio frequency interference
RMS	root mean square
TL	transmission line
VSI	voltage source inverter

Glossary

Adjustable speed drive (ASD) A converter using controlled rectifier or transistor devices that has the capability of adjusting the frequency and proportional voltage of the output waveform to provide speed control of motors. (IEEE Std. 45-2002)

Asymmetrical terminal voltage Common mode voltage, measured by means of a delta network at specified terminals. (IEC 60050-161-04)

Attenuation A general term used to denote a decrease in signal magnitude in transmission from one point to another. (IEEE Std. 1050-2004)

Broadband electrical noise (interference) Electrical noise that contains energy covering a wide frequency range. (IEEE Std. 1143-1994)

Carrier frequency In pulse-width-modulated (PWM) switching schemes, the switching frequency that establishes the frequency at which the converter switches are switched. In sine-triangle PWM, the carrier frequency is the frequency of the triangle waveform that the control or modulating signal is compared to. (Ed. PA Laplante, *Comprehensive Dictionary of Electrical Engineering*, 2nd edition, CRC Press, 2000)

Broadband interference (disturbance) (1) An undesired emission that has a spectral energy distribution sufficiently broad so that the response of the measuring receiver in use does not vary more than 3 dB when tuned over the frequency range of plus or minus two impulse bandwidths. (ANSI C63.14-2009)
(2) An electromagnetic disturbance that has a bandwidth greater than that of a particular measuring apparatus, receiver, or susceptible device. For some purposes, particular spectral components of a broadband disturbance may be considered as narrowband disturbances. (IEC 60050-161-06)

Characteristic impedance The ratio of the complex voltage and complex current of a signal traveling forward on a conductive path. A signal path is often terminated with an impedance that matches the characteristic impedance of the path. This makes the path appear to be infinitely long and prevents signal degradation due to reflections that occur at unterminated ends of the path. (IEEE Std. 1149.6-2003)

Coupling The association of two or more circuits or systems in such a way that power or signal information may be transferred from one system or circuit to another. (IEEE Std. 1159-2009)

Coupling path The path over which part or all of the electromagnetic energy from a specified source is transferred to another circuit or device. (IEC 60050-161-03)

Coupling (electric, capacitive) Electric field coupling between a source and a grounded circuit modeled by a capacitor. Capacitive coupling occurs when a varying electrical field exists between two adjacent conductors typically less than a wavelength apart, inducing a change in voltage across the gap. (B. Zhang and X. Wang, *Chaos Analysis and Chaotic EMI Suppression of DCDC Converters*, John Wiley & Sons, Inc., 2015)

Coupling (magnetic, inductive) Magnetic field coupling between a current carrying conductor and another conductor modeled by a transformer. Magnetic coupling occurs when a varying magnetic field exists between two parallel conductors typically less than a wavelength apart, inducing a change in voltage along the receiving conductor. (B. Zhang and X. Wang, *Chaos Analysis and Chaotic EMI Suppression of DC-DC Converters*, John Wiley & Sons, Inc., 2015)

Crosstalk An undesired signal disturbance introduced in a transmission circuit by mutual electric (capacitive) or magnetic (inductive) field coupling with other transmission circuits. (ANSI C63.14-2014)

Converter A generic term used in the area of power electronics to describe a rectifier, inverter, or other power electronic device that transforms electrical power from one frequency and voltage to another. (Ed. PA Laplante, *Comprehensive Dictionary of Electrical Engineering*, 2nd edition, CRC Press, 2000)

Conducted emission (1) An RF current propagated through an electrical conductor. (Ed. PA Laplante, *Comprehensive Dictionary of Electrical Engineering*, 2nd edition, CRC Press, 2000)
(2) Electromagnetic emissions propagated along a metallic conductor, which could be a power line, signal line, and/or an unintentional or fortuitous conductor, such as a metallic pipe, and so on. (ANSI C63.14-2014)

Conducted emission frequency range Frequency from 9 kHz up to 30 MHz. (ANSI C63.12-2015)

Conducted disturbance Electromagnetic disturbance for which the energy is transferred via one or more conductors. (IEC 60050-161-03)

Conducted interference (disturbance) Undesired electromagnetic energy that is propagated along a conductor, usually defined in terms of a voltage and/or current level. (ANSI C63.14-2014)

Conducted noise Unwanted electrical signals that can be generated by power electronic switching circuits. Conducted noise can travel through the circuit cables as common-mode or differential mode currents and can interfere with control circuits or other electronic equipment. (Ed. PA Laplante, *Comprehensive Dictionary of Electrical Engineering*, 2nd edition, CRC Press, 2000)

Common mode impedance The quotient of the common mode voltage by the common mode current. (IEC 60050-161-04)

Common mode voltage (CMV) (asymmetrical voltage) The mean of the phasor voltages appearing between each conductor and a specified reference, usually earth or frame. (IEC 60050-161-04)

Common mode voltage (CMV) In the context of adjustable speed drives (ASDs), CMV is the displacement of the neutral point (and each phase voltage) of the ASD output from ground due to the switching of the solid-state devices in the drive. It is an alternating voltage whose magnitude and frequency components are dependent on the drive topology. All present drive topologies create CMV to some extent. CMV can also be created at the motor if phase circuit conductors, unsymmetrical with respect to the equipment grounding conductor(s) or grounded sheaths or raceways, are used between the ASD output and the motor. (IEEE Std. 1349-2011)

Common mode (CM) current In a cable having more than one conductor, including shields and screens if any, the magnitude of the sum of the phasors representing the currents in each conductor. (IEC 60050-161-04)

Common mode (CM) circuit The full current loop or closed circuit for the CM current, including the cable, the apparatus, and the nearby parts of the earthing system. (ANSI C63.14-2014)

Common mode (CM) conversion The process by which a differential mode voltage is produced in response to a common mode voltage. (IEC 60050-161-04)

Current probe A device for measuring the current in a conductor without interrupting the conductor and without introducing significant impedance into the associated circuits. (IEC 60050-161-04)

DC link The coupling between the rectifier and inverter in a variable speed AC drive. (Ed. PA Laplante, *Comprehensive Dictionary of Electrical Engineering*, 2nd edition, CRC Press, 2000)

DC link capacitor A device used on the output of a rectifier to create an approximately constant DC voltage for the input to the inverter of a variable speed AC drive. (Ed. PA Laplante, *Comprehensive Dictionary of Electrical Engineering*, 2nd edition, CRC Press, 2000)

DC link inductor An inductor used on the output of a controlled rectifier in AC current source drives to provide filtering of the input current to the current source inverter. If used in conjunction with a capacitor, then it is used as a filter in voltage source drives. (Ed. PA Laplante, *Comprehensive Dictionary of Electrical Engineering*, 2nd edition, CRC Press, 2000)

Differential mode voltage (symmetrical voltage) The voltage between any two of a specified set of active conductors. (IEC 60050-161-04)

Differential mode (DM) circuit The full current loop or closed circuit for the intended signal or power, including a cable and the apparatus connected to it at both ends. *Synonym:* symmetric-mode circuit. (ANSI C63.4-2014)

Differential mode current In a two-conductor cable, or for two particular conductors in a multiconductor cable, half the magnitude of the difference of the phasors representing the currents in each conductor. (IEC 60050-161-04)

Electromagnetic compatibility (EMC) (1) The ability of an equipment or a system to function satisfactorily in its electromagnetic environment without introducing intolerable electromagnetic disturbances to anything in that environment. (IEC 60050-161-01)

(2) The ability of systems, equipment, and devices that utilize the EM spectrum to operate in their intended operational environments without suffering unacceptable degradation or causing unintentional degradation because of EM radiation or response. It involves the application of sound EM spectrum management; system, equipment, and device design configuration that ensures interference-free operation; and clear concepts and doctrines that maximize operational effectiveness. (C63.14-2014)

Electromagnetic disturbance (EMD) (1) Any electromagnetic phenomenon that can degrade the performance of a device, an equipment, or a system, or adversely affect living or inert matter (IEC 60050-161-01). An electromagnetic disturbance can be an electromagnetic noise, an unwanted signal, or a change in the propagation medium itself.

(2) Any EM phenomenon that may degrade the performance of a device, equipment, or system, or adversely affect living or inert matter (C63.14-2014). An EM disturbance may be a noise, an unwanted signal, or a change in the propagation medium. (ANSI C63.4-2009)

Electromagnetic interference (EMI) (1) Any electromagnetic disturbance that interrupts, obstructs, or otherwise degrades or limits the effective performance of electronics/electrical equipment. It can be induced intentionally, as in some forms of electronic warfare, or unintentionally, as a result of a spurious emissions and responses, intermodulation products, and the like. Additionally, EMI may be caused by atmospheric phenomena, such as lightning and precipitation static and non-telecommunication equipment, such as vehicles and industry machinery. (Ed. PA Laplante, *Comprehensive Dictionary of Electrical Engineering*, 2nd edition, CRC Press, 2000)

(2) Unwanted high-frequency electrical signals, also known as radio frequency interference (RFI), which can be generated by power electronic

circuits switching at high frequencies. The signals can be transmitted by conduction along cables or by radiation and can interfere with control or other electronic equipment. (Ed. PA Laplante, *Comprehensive Dictionary of Electrical Engineering*, 2nd edition, CRC Press, 2000)

(3) Degradation of the performance of an equipment, a transmission channel or a system caused by an electromagnetic disturbance.
In French, the terms perturbation lectromagntique and brouillage lectromagntique designate, respectively, the cause and the effect, and should not be used indiscriminately. In English, the terms electromagnetic disturbance and electromagnetic interference designate, respectively, the cause and the effect, but they are often used indiscriminately. (IEC 60050-161-01)

EMI filter Electromagnetic interference filter used to reduce or eliminate the electromagnetic interference (EMI) generated by the harmonic current injected back onto the input power bus by switching circuits. The harmonic current is caused by the switch action that generates switch frequency ripple, voltage and current spikes, and high-frequency ringing.

EMI measurement frequency bands CISPR publication 16 splits the EMI measurement frequency range of 9 kHz to 1 GHz into four bands: band A (9–150 kHz) and band B (150–30 MHz) for conducted EMI, band C (30–300 MHz) and band D (300–1 GHz) for radiated EMI, in which different time constants and resolution bandwidths of EMI receiver are used. (CISPR 16-1-1)

EMI receiver (1) A measuring instrument that meets the requirements called out in ANSI C63.2-1996 or CISPR 16-1-1:2006.
(2) scanning receiver is a measuring instrument that complies with the requirements called out in ANSI C63.2-1996 or CISPR 16-1-1:2006, as applicable, and does not cover a frequency range in discrete frequency steps but continuously tunes across it. (ANSI C63.14-2014)

Emission (electromagnetic) (1) Electromagnetic energy propagated from a source by radiation or conduction. (ANSI C63.14-2014)
(2) The phenomenon by which electromagnetic energy emanates from a source. (IEC 60050-161-01)

Emission spectrum The distribution of the amplitude (and sometimes phase) of the components of an emission as a function of frequency. (ANSI C63.14-2014)

Frequency converter A machine, device, or system for changing AC at one frequency to AC at a different frequency. (IEEE Std. 388-1992)

Ground (GND) (earth) (1) The conductive mass of the earth, which has an electric potential at any point that is conventionally taken as equal to zero.
(2) A conducting connection, whether intentional or accidental, between an electrical circuit and the earth, or to some conducting body that serves in place of earth.

(3) The position or portion of an electrical circuit at zero potential with respect to the earth.

(4) A conduction body, such as the earth or the hull of a steel ship, used as a return path for electric currents and as an arbitrary zero reference point. *Synonym:* earth.

(5) A continuous metal-to-metal bond, existing or provided, around the outer perimeter of a metallic item or cable shield terminating at or penetrating through a metal surface that is at ground potential. (ANSI C63.14-2014)

High frequency (HF) (in electromagnetic compatibility) Frequency above 9 kHz. (IEC 60050-161-03)

Insertion loss The ratio of voltages, current, or power at a given frequency, appearing across the line immediately beyond the point of insertion, before and after insertion of the filter under test. (IEEE Std. 1560-2005)

Line impedance stabilization network (LISN) A network inserted in the power supply lead of apparatus to be tested that provides, in a given frequency range, a specified load impedance for each current carrying conductor for the measurement of disturbance voltages. LISN may isolate the apparatus from the supply mains in that frequency range as well as couple the EUT emissions to the measuring instrument. (ANSI C63.14-2014)

Low frequency (LF) (in electromagnetic compatibility) Frequency up to and including 9 kHz. (IEC 60050-161-03)

Lumped element A circuit element of inductors, capacitors, and resistors; its dimension is negligible relative to the wavelength. (Ed. PA Laplante, *Comprehensive Dictionary of Electrical Engineering*, 2nd edition, CRC Press, 2000)

Parasitics Electrical properties of a design (resistance, capacitance, and impedance) that arise due to the nature of the materials used to implement the design. (IEEE STD 1481-2009)

Parasitic capacitance The generally undesirable and not-designed-for capacitance between two conductors in proximity of one another. (Ed. PA Laplante, *Comprehensive Dictionary of Electrical Engineering*, 2nd edition, CRC Press, 2000)

Parasitic inductance The generally undesirable and not-designed-for inductance associated with a conductor, or path of current on a conductor. (Ed. PA Laplante, *Comprehensive Dictionary of Electrical Engineering*, 2nd edition, CRC Press, 2000)

Power grid An assembly of normally interconnected power systems arranged to meet the power generation and consumption needs of a relatively large geographic area. (IEEE Std 1826-2012)

Propagation velocity The velocity at which an electric signal travels through a cable. Propagation velocity is usually expressed in feet, yards, or meters per microsecond or as a percentage of the speed of light. The value of the propagation velocity depends on the (relative) dielectric constant of the insulation material used, the characteristic of the semicon shields, and the cable construction; it is assumed constant for all practical purposes. (IEEE Std 1234-2007)

Radiation The phenomenon by which sources generate energy, which propagates away from them in the form of waves. (Ed. PA Laplante, *Comprehensive Dictionary of Electrical Engineering*, 2nd edition, CRC Press, 2000)

Radiated emission (1) An electromagnetic field propagated through space. (Ed. PA Laplante, *Comprehensive Dictionary of Electrical Engineering*, 2nd edition, CRC Press, 2000)
(2) Desired or undesired electromagnetic energy, in the form of electric and magnetic fields, which is propagated through space. (ANSI C63.14-2014)

Radio frequency interference (RFI) Electromagnetic phenomenon that either directly or indirectly contributes to degradation in the performance of a receiver or other RF system, synonymous with electromagnetic interference. (Comprehensive Dictionary of Electrical Engineering, Ed. P.A. Laplante, CRC Press, 2000)

Radiated emission frequency range Frequency above 30 MHz (ANSI C63.12-2015)

Radiated disturbance Electromagnetic disturbance for which the energy is transferred through space in the form of electromagnetic waves. The term "radiated disturbance" is sometimes used to cover induction phenomena. (IEC 60050-161-03)

Reference-ground plane (RGP), reference-earth plane Flat conductive surface that is at the same electric potential as reference ground, which is used as a common reference and which contributes to a reproducible parasitic capacitance with the surroundings of the equipment under test (EUT). A reference-ground plane is needed for the measurements of conducted disturbances, and serves as reference for the measurement of unsymmetrical and asymmetrical disturbance voltages. In some regions, the term earth is used in place of ground. (IEC 60050-161-04)

Screen A device used to reduce the penetration of a field into an assigned region. (IEC 60050-161-04)

Screen (electromagnetic) A screen of conductive material intended to reduce the penetration of a varying electromagnetic field into an assigned region. (IEC 60050-161-04)

Shield (Instrumentation cables) (cable systems) A metallic sheath, usually copper or aluminum, applied over the insulation of a conductor or conductors for the purpose of providing means for reducing electrostatic coupling between the conductors so shielded and others that may be susceptible to or that may be generating unwanted (noise) electrostatic fields. (IEEE Std. 1250-2011)

Shielded cable A cable in which each insulated conductor or conductors is/are enclosed in a conducting envelope(s). (IEEE Std. 1234-2007)

Symmetrical terminal voltage Differential mode voltage, measured by means of a delta network at specified terminals. (IEC 60050-161-04)

Termination An impedance usually near the end of a signal path used to satisfy the electrical matching requirements of the characteristic impedance of the signal path and reduce signal reflections. (IEEE Std. 1149.6-2003)

Transfer impedance (of a screened circuit) The quotient of the voltage appearing between two specified points in the screened circuit by the current in a defined cross section of the screen. (IEC 60050-161-04)

Transmission line Typically, a uniform conductor pair, forming a continuous path from an electrical energy source to a receptor, for directing (conducting) the transmission of electromagnetic energy along this path. In practice, typical transmission line configurations include telephone lines, power cables, coaxial cables, and computer cables. (ANSI C63.14-2014)

1

Introduction to Conducted Emission in Adjustable Speed Drives

> *It is never possible to predict a physical occurrence with unlimited precision.*
>
> Max Planck

1.1 High-Frequency Emission of Switch-Mode Power Converters

Switch-mode power conversion method is generally based on the switching regulator that allows controlling output voltage and current by changing the ratio between on-time and off-time of switches by using different modulation patterns. The efficiency of switch-mode power conversion depends mostly on switching characteristics of semiconductor switches. Reducing of transistors' switching times allows decreasing switching losses and increasing carrier frequency of modulation [1,2]. Faster switching of switches in switch-mode power converters, although very beneficial for energy conversion efficiency, unfortunately results in faster voltage and current changes [3,4].

Any rapid change of voltage or current in electric circuit, from the EMC point of view, is a source of electromagnetic emission that can be potentially harmful to other equipment. Therefore, these emissions should be limited to provide undisturbed operation of electric devices. Electromagnetic emission spectrum width, which directly depends on the rate of change of voltages and currents, in currently used power electronic devices can easily achieve megahertz band [5,6].

With the increase of frequency of electromagnetic emission, effectiveness of its unintentional propagation capabilities usually increase, thus enabling the generated disturbances to be transmitted more easily by means of conduction

High Frequency Conducted Emission in AC Motor Drives Fed by Frequency Converters: Sources and Propagation Paths, First Edition. Jaroslaw Luszcz.

or radiation phenomena. To reduce electromagnetic emission of switch-mode power converters, wide-band passive filters and shields are most often used. Design and implementation of EMI filtering components are troublesome and expensive, especially in low frequency (LF) range [7–9], below 9 kHz, where sizes of filtering components are considerable, as well as in high frequency (HF) range, above 9 kHz, where radiated and conducted emission leakages are very difficult to predict and avoid at the design stage [10,11].

Furthermore, the reduction of electromagnetic emission usually becomes more difficult with the increase of converter's rated power because of the overall size of converter's components that in turn results in increase of parasitic couplings, whose effects are more difficult to avoid or even to decrease [12–14].

1.2 Characteristic Issues of Conducted Emission in Adjustable Speed AC Motor Drives

ASDs consisting of AC motors and FC containing voltage source inverters (VSI) controlled according to pulse width modulation (PWM) patterns are exceptional in several ways in relation to many other types of switch-mode high-power converters integrated with power systems.

First, in recent years the total power of ASD used in residential, commercial, and industrial environments has been increasing significantly. This considerable increase is observed primarily in rapidly growing number of installed low-power ASDs, below few kilowatts, in critical environments containing electromagnetically sensitive equipment, for example, air-conditioning systems, intelligent and automated buildings, fully automated production lines, and energy-saving installations. On the other hand, the rated power of single ASD used in heavy industry and directly integrated with the power system at distribution level achieves quite often rated power even in megawatts. ASD of high rated power of megawatts are more often powered directly from medium voltage (MV) grids, which results in commutation of much higher DC voltages and leads to increased generation of conducted emission [15–17].

Second, currently used semiconductor switches, mostly IGBT transistors of different generations, commutate high voltages and high currents in shorter time in order to decrease switching losses and allow using higher modulation carrier frequencies. High-power IGBT transistors used in high-power frequency converters exhibit relatively large parasitic capacitances between semiconductor substrate and cooling subbase. Higher levels of switched voltages together with shorter switching times and more considerable parasitic couplings result in significant increase of transient capacitive stray currents flowing between energized and grounded components, which are essential for conducted emission generation and propagation [18,19].

Third, load of FC—AC motor windings together with motor feeding cable—cannot be precisely taken into account at the design stage of FC, because it is widely dependent on requirements of singular application. Thus, HF parameters of FC load can be known only at installation stage and can usually differ significantly in each particular application of the same type of FC [20,21]. The physical size of FC load, feeding cable with windings, is usually much longer compared to distances encountered inside FC and in many applications it can obtain the length comparable to the wavelength of transmitted harmonic components of signals. In such applications, more effective propagation of generated conducted emission toward other adjacent installations and systems can occur. Some manufacturers publish its own recommendations for installation of their frequency converters, which specify the suggested configuration of converters' output circuit that allow avoiding EMC violation problems. Majority of these recommendations is related to motor feeding cable specification and wiring style, especially grounding connections.

The problem is that the length of motor cable and its arrangement depends substantially on requirements of particular application and cannot be precisely predicted or fixed at the design stage of ASD. Particular parameters of motor cable can influence significantly overall EMC performance of ASD, which can necessitate using an extra filtering technology at the output and input sides of FC [21,22]. Some typical filtering solutions correlated with the length of motor cable are also recommended by manufacturers [23,24]. Unfortunately, even strict accordance to manufacturer's recommendations quite often is not sufficient to ensure lack of interference in electrical systems containing ASD. Usually each ASD installation requires the use of additional measures to maintain generated conducted emission at acceptable level.

The most frequently occurring EMC problems in control systems with ASD are usually associated with

- high levels of common mode (CM) currents at converter's output caused by wire-to-ground parasitic capacitances,
- overvoltage transients at motor terminals as a result of impedance mismatch,
- excessive transfer of CM currents from converter's output side toward power grid,
- high AC motor internal CM currents damaging motor bearings,
- stray HF current circulating in surrounding grounded components as a potential source of interfering effect for other systems,
- high levels of radiated emission close to motor cable route, easily coupled to other systems, and
- significantly elevated narrowband conducted and radiated emission within selected frequency bands as an effect of motor cable length.

Intensity of these effects is difficult to predict accurately using known procedures of radio frequency interference (RFI) filters design recommended for converters' load side, and it is therefore hard to avoid. There are number of issues that make procedures used for designing RFI filtering at the output side of FC particularly difficult. The most significant of them are related to

- insufficient standard recommendations related to converters' output side, motor windings, and motor cables;
- difficulties with accurate determination of parameters for models of AC motor windings and motor feeding cable in a wide frequency range;
- lack of manufacturer's specification and severe difficulty with experimental determination of parasitic parameters of frequency inverters—especially internal stray capacitances between the energized components and ground; and
- possible resonance interactions between motor windings, the feeding cable, and output filters in HF range.

Fourth, conducted emission at the output side of frequency converter of ASD is not directly limited by current standard recommendations. If the grid-side conducted emission of ASD is maintained within the required limits, the output-side filtering is very often applied only in case of appearance of EMC problems in installation at the commissioning stage, very rarely at the design stage. It is a result of the fact that there are a number of ASD applications functioning successfully without any output filtering applied [25,26]. In contrast to other applications with power electronics converters, in ASD motor cables connected to output sides of frequency converters are very often placed along the same cable trays as other power cables. Such arrangement of motor cables results in the creation a propagation path that enables efficient coupling of output-side-originated conducted emission of frequency converter directly toward power system with bypassing grid-side filters of the converter [27,28]. Apart from conducted emission that can be injected by the ASD motor feeding cable directly into power grid by means of crosstalk phenomena between nearby power cables, there is also often encountered interference in nearby placed control cables, including connections of the speed control system of the ASD itself [29–31].

Fifth, in contemporary power grids, there are already many situations where a significant number of ASDs are used and this trend is increasing continuously. In such electromagnetic environment with high levels of electromagnetic emission generated by ASD, interference issues more often occur. Future power grids are expected to be developed toward smart grids in which advanced measurement and control technologies will be used and therefore the significance of electromagnetic emission and immunity levels of electrical devices in systems is also expected to increase [32–34]. The foreseen increase of threats of EMC in the future electrical systems may be associated with at least two trends already observed [35,36]:

- More electric power is converted using static converters, including ASD, at different levels of power grid: generation, transmission, and distribution, which will presumably increase overall levels of harmonic emission, even if standardized limitation for individual devices will be fulfilled [37–39],
- The increasing quantity of measuring and communication devices are widely used in order to monitor and control optimal energy-saving power flow, which can be harmfully influenced by electromagnetic disturbances originating from neighboring high-power static converters [40,41].

1.3   Essential EMC Problems of Integration of ASD with the Power Grid

Contemporary problems with integration of ASD with the power grid (PG) are predominantly related to its unintentional electromagnetic emission in various frequency bands [42,43]. EMD, as any electromagnetic phenomenon that may negatively influence the performance of other devices, are generated by any voltages and currents varying in time. Its level and spectrum content depend on the magnitude of changing voltage and the speed of change. In the last decade, the speed of change of signals occurring typically in ASD increased significantly due to successful development of power electronics transistor technology that allows switching higher currents and higher voltages in shorter time [44,45]. These capabilities result positively in significant increase of converted power at higher efficiency, but unfortunately also negatively in higher electromagnetic emission in a wider frequency range.

In ASD, electromagnetic disturbances are generated by different components and therefore variously and specifically distributed over the frequency spectrum. EMD, depending on its spectral content, can influence other devices and systems in different ways. First, EMD propagation paths are strongly related to its spectral content and, second, interfered device's susceptibilities highly depend on frequency of disturbing signals [5,46]. Over the years, many standards have developed specifications of electromagnetic emission and susceptibility characteristics for different devices in specific frequency ranges that are tightly correlated with particular phenomena that may disturb other devices [47–50].

According to the current standard definitions, electromagnetic emissions are usually classified into four categories based on its two characteristic frequency range: 9 kHz and 30 MHz. The frequency range above 9 kHz has been well established since long time as radio frequencies (RF) and most of EMC issues localized primarily in this range are named as electromagnetic interference (EMI). The frequency 9 kHz is also defined by IEC 60050-161 standard as frequency limit for low frequency (LF) band and high frequency (HF) band

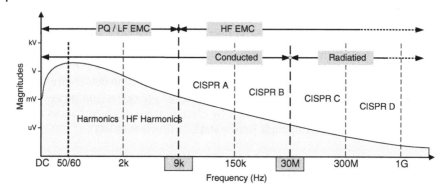

Figure 1.1 Electromagnetic emission frequency bands defined by PQ and EMC related standards.

recommended for categorization of phenomena in the field of electromagnetic compatibility (Figure 1.1).

The second limit of 30 MHz categorizes EMC phenomena for conducted emission below 30 MHz and radiated emission above 30 MHz. This split is based on the assumption that in majority of typical cases of unintended emission for frequencies lower than 30 MHz, the predominant part of electromagnetic energy is transferred by conduction via cable connections or other conductive components. Above 30 MHz, the dominating part of electromagnetic energy is propagated through space in the form of electromagnetic waves (Figure 1.1). EMD and EMI issues in frequency range below 9 kHz are usually called as low-frequency EMC phenomena with particular emphasis on harmonic distortions below 2 kHz, specified by total harmonic distortion (THD) factor that is one of power quality (PQ) index. In the frequency range between 2 and 9 kHz, the method of grouping harmonic distortion within 200 Hz wide subbands is already proposed in some standards related to high-frequency harmonic distortion limitations in power grids, for example, IEC 61000-4-7.

In general, despite some specific cases, magnitudes of EMD generated by typical power electronic applications decrease with frequency (Figure 1.1), starting from several percent of nominal RMS values of grid voltage or current within frequency range close to the grid frequency and reaching much smaller levels of only microvolts or microamperes for upper frequency range of conducted EMI band, close to 30 MHz. Unfortunately, even so low voltage and current amplitudes can be actually harmful, disturbing, and difficult to eliminate because of relatively high frequency that results in easy propagation by means of omnipresent parasitic couplings [51,52].

Currently, ASD applications are considered as one of the most disturbing sources of electromagnetic emission in a wide frequency range, from frequency of power grid up to several of megahertz. Wide spectrum and high levels of EMD

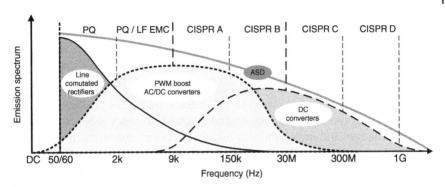

Figure 1.2 Characteristic emission spectra of typical power electronic converters commonly used in contemporary applications.

emission require to use multiple filtering techniques applied simultaneously in individual frequency subbands to reduce emission in troublesome applications to the levels accepted by standards or immunity of particular applications.

First, grid-side AC/DC converters used in ASD of low and medium rated power are in majority diode-based rectifiers. High levels of harmonic distortions in frequency range of up to 2 kHz by diode rectifiers, are well known as one of the reasons of power quality (PQ) degradation in power grids. To reduce that problem, controlled rectifies are introduced, which are based on thyristor or transistor technology. They allow limiting efficiently harmonics emission in frequency range below 2 kHz, although they simultaneously increase significantly emission in higher frequency ranges, up to few hundreds of kilohertz [53,54]. This technology is used in ASD as PWM boost rectifiers for individual drives or common DC bus ASD and as active filtering for multiple nonlinear loads. Simplified spectral differences in emission characteristics of different types of static converters are presented in Figure 1.2.

Second, ASD's output inverter, usually IGBT transistor based, generate high levels of EMD similar to DC/DC converters in HF conducted EMD frequency range of 9 kHz up to 30 MHz that can easily propagate toward the power grid through internal DC bus connections. This conducted emission can be limited using filters for radio frequency interference (RFI) [54–56].

Third, the load of FC in ASD is very exceptional in relation to classic DC/DC power electronic converters. It usually consists of extraordinary long motor feeding cable connected directly to converter terminals and thus can significantly influence the resultant ASD emission. There are two specific and critical effects of high-level EMD generated at output side of FC that are difficult to eliminate for solving EMC problems in applications with ASD:

- High levels of radiated emission nearby motor side of frequency converter, AC motor, and along motor cable that pollute electromagnetic environment and can be easily coupled to nearby located systems by means of near-field couplings [57–59].

- Increased conducted emission at grid side of FC caused by transfer of conducted emission generated at motor side of FC to the power grid through DC bus connection and parasitic couplings inside FC [54,60,61].

Both of these effects are related to HF voltage transients generated at converters output terminals, in motor windings, and in motor cable due to rapid voltage changes and impedance mismatches between converter, motor cable, and motor windings.

Finally, as it is shown in Figure 1.2, ASDs can generate significant levels of EMD in frequency ranges, primarily associated with different types of static converters, that can also be extra increased because of interactions of converter output with relatively large-scale external output circuitry, motor cable, and motor windings. Precise prediction of those HF interactions is difficult at converter's design stage, because broadband parameters of motor winding and motor cable can vary significantly in each particular application. Type of motor, type of feeding cable, and especially motor cable length and its mutual arrangement in relation to other components can be the most decisive factors.

Most of EMC and PQ issues appearing in an ASD can be divided into two categories: related to meeting of EMC standards or other regulations and directly resulting from susceptibility to other systems. Standard regulations are focused mainly toward protection of power grid against injection of excessive conducted emission and preservation of electromagnetic environment against radiated emission levels that can be harmful for electrical equipment and also living creatures. These problems are usually solved by using power lines filtering methods used in radio frequency range and device components shielding, often with significant efforts. Problems with internal EMC are usually much more challenging and require to minimize generation of EMD at its source that is essential for devices or systems themselves, because it allows minimizing external filtering and shielding demand.

This approach can be particularly effective for solving EMC problems in an ASD because internal compatibility of systems that include ASDs are often critical and also its external emission can be considerably limited by lowering internal emissions.

Essential part of the ASDs that is very critical for its internal and external EMC is FC load circuit that can be very influential and cannot be specified in details by the FC manufacturer. The FC load, such as AC motor windings and feeding cable, has to be configured individually by the ASD end user and can change significantly its final emission characteristic. There are some helpful recommendations known for designing inverter's output circuits delivered by FC manufacturers; nevertheless, application of these principles is sometimes difficult and cannot entirely guarantee the avoidance of all possible negative effects of FC load. Particular application requirements, which result from variable cable length, its layout, and parameters of motor winding and cable can change EMD emission of an ASD significantly.

Motor feeding cable and its interactions with motor windings in frequency range of megahertz can induce negative effects, for example, elevated conducted and radiated emission, excessive CM currents flow, bearing currents, and elevated exposure of motor windings to overvoltages [62–64].

Nowadays there are many investigations carried out in order to find efficient modeling method that allows estimating potential EMC problems in ASD application at the design stage. Effectiveness of wide-band modeling of ASD is essentially related to the adequacy of used models of all system components. Wide-band models of typical ASD components, adequate in conducted EMI frequency range, are usually relatively complex, because they require to take into account parasitic couplings that cannot be neglected in high-frequency bands. More accurate models of ASD components usually require much more efforts for identification of its parameters, so the main problem in wide-band modeling is the trade between model adequacy and its parameters identification overheads. On the other hand, the use of simplified models usually limits significantly a frequency range within which its adequacy can be accepted [65,66].

1.4 Scope of the Book

The objective of this monograph is to present the state-of-the-art analysis of undesirable high-frequency phenomena accompanying AC motor speed control by using voltage source inverter with PWM modulation pattern. The major part of this book is focused on the proposed new approach for wide-band analysis of AC motor with feeding cable, using simplified circuit models in frequency range of conducted emission up to 30 MHz. Theoretical analyses presented in this book have been compared with results obtained by experimental investigations carried out for selected types of low-power ASD applications. The new methods proposed for wide-band analysis of output circuit of frequency inverter in wide frequency range are highly focused on reasonable balance between modeling adequacy, model complexity with its parameters identification efforts, and computational overheads of simulation analysis.

In Chapters 1 and 2, the sources of high levels of conducted emission generated in ASDs are identified. In these chapters the fundamentals of EMI generation phenomena in switch-mode static converters are described and discussed. Analytical and graphical methods of emission spectra estimation are discussed based on the comparison of results obtained analytically with those obtained experimentally. The problem of model structure complexity and difficulties in identification of its parameters is underlined in relation to the achieved accuracy of attained results.

In Chapters 3 and 4, propagation phenomena of EMI generated by FC into the power grid are described and explained based on theoretical background and results obtained experimentally for exemplary low-power ASDs. Crucial parasitic capacitive couplings as an essential propagation mechanism are specified. In these chapters, the key impact of EMI generated at output side of frequency converter on input-side common mode currents injected by an ASD into the power grid is underlined. Parasitic capacitances of internal DC bus link in FC are pointed out as a foremost EMI propagation pathway between converter's output side and grid side.

In Chapters 5 and 6, methodology of modeling conducted emission generation and propagation in ASDs is presented. In these chapters, new simplified wide-band models of AC motor windings together with motor feeding cable are proposed and verified by simulation analysis and experimental investigation. New approach, based on time domain transient analysis, is proposed and discussed in relation to other methods that are relatively more complex, require more laborious procedures of model's parameters identification, and result in not very reliable final effects.

Finally, in Chapter 7, the influence of wide-band behavior of FC load, AC motor windings, and motor feeding cable on grid-side conducted EMI emission levels are analyzed theoretically using the proposed models. The obtained simulation results are compared with those obtained experimentally for the evaluated, typically used low-power ASDs with a diode rectifier as grid-side AC–DC converter. In this chapter, the applicability of the proposed method for estimating the influence of AC motor and feeding cable parameters on the overall EMC performance of the ASD is presented.

1.5   EMC-in-ASD Related Assumptions, Definitions, and Naming Conventions

Nowadays many different types and configurations of ASDs are in use. The used technologies include voltage or current source inverters, carrier PWM, or more advanced modulation strategies [67,68] with six or more pulse diode rectifiers as an AC–DC input converter [69], PWM boost AC–DC input converter [70], two levels or multilevel output inverters [71,72], integrated input and/or output LF filters and RFI filters [73,74], different motor winding configurations [75], various motor feeding cables [76,77], and different power electronics switches [78]. With the more advanced technology with more optional components applied, for example, RFI filters, the analysis of EMC issues in ASD becomes more and more complex. However, it is rather difficult to analyze EMC in detail in even relatively simple circuits.

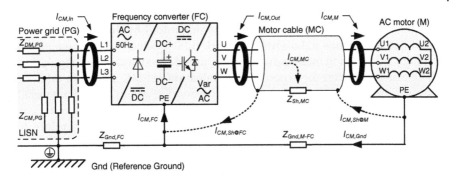

Figure 1.3 General configuration of grid-connected ASD and EMC-related naming convention.

Investigations presented in this book are primarily concentrated on EMC-related phenomena that take place at output side of frequency converter. Therefore, the configuration of the evaluated ASD has been chosen with preferences for maximum simplification of analyzed matter, especially those that are not directly related to investigated issues. The applied simplifications are focused mainly on elimination of external filtering components and using as simple AC–DC grid-side converter as possible, to avoid interactions in HF range that can significantly interfere investigated issues.

Theoretical analysis and experimental test presented in this book have been done for relatively simple ASD system consisting of

- three-phase full-wave diode-based rectifier,
- VSI inverter build as three-phase bridge based on integrated IGBT module,
- externally accessible DC link bus with capacitive and inductive filtering,
- three-wire-shielded motor feeding cable, and
- low-voltage four-pole AC motor with star-connected windings and floating star point.

The general configuration of evaluated ASD is presented in Figure 1.3 where the most significant details associated with its EMC performance are specified. The power grid (PG) is usually modeled by differential mode impedances $Z_{DM,PG}$ and common mode impedances $Z_{CM,PG}$ that represent line-to-line and line-to-ground frequency-dependent impedance characteristics seen from ASD input terminals. Unfortunately, values of these impedances are difficult to identify thoroughly, especially in a wide frequency range because they

- depend significantly on frequency band,
- depend on the localization of connection point to power grid,
- are time variable due to the power system load fluctuations, and
- can be influenced by loads connected locally to the power system.

The troublesome influence of the power grid impedance variations, which can change measurement results, can be minimized by the use of standardized line impedance stabilization network (LISN), which is recommended by conducted emission measurement procedures defined in standard regulations. The essential matter for experimental evaluation of EMC issues is a reference ground. In ASD it is a problematic issue because FC and AC motor are usually bounded to ground; nevertheless, these bonding connections cannot be considered as zero impedance connections, especially in HF range. Therefore, for presented analysis, the main grounding point of LISN was used as reference ground, despite that the impedance of ground connection of LISN to FC has been minimized efficiently in the whole frequency band of conducted EMI.

Reliable grounding of ASD components—like FC, AC motor, and motor feeding cable shield—is required primarily for safety reasons; this requires sufficient current capacity to prevent the buildup of hazardous voltages at all accessible conducting parts of installation. The impedance characteristic of grounding connections in HF range is one of the most important factors influencing reduction of conducted and radiated emission of ASD. Therefore, grounding connections, especially between FC and AC motor, should be designed to minimize impedances of grounding loops in HF range. One of the possible effective solutions is the use of shielded motor feeding cables with purposely designed shield structure with added extra returning wires to achieve beneficial impedance characteristic within the widest range of frequencies.

The majority of analysis presented in this book are focused primarily on CM currents distribution in ASD; therefore, a number of CM currents and impedances categories associated with them—presented in Figure 1.3—are defined as follows:

$[I_{CM,In}]$	input side CM current of FC that is injected into the power grid by ASD
$[I_{CM,Out}]$	output side CM current of FC that enters motor feeding cable
$[I_{CM,M}]$	motor CM currents that flow through parasitic capacitances of motor windings toward the grounded stator
$[I_{CM,MC}]$	CM current of motor cable, a part of total CM output current of inverter that flows through parasitic capacitances of motor feeding cable wires and shield
$[I_{CM,ShM}]$	CM current returning back from motor to shield of motor feeding cable
$[I_{CM,Gnd}]$	CM current returning back from motor to FC by ground connections other than motor cable shield

$[I_{CM,ShFC}]$	CM current returning back to FC by motor feeding cable shield
$[I_{CM,FC}]$	total CM current of FC
$[Z_{CM,PG}]$	CM impedances of power grid
$[Z_{DM,PG}]$	DM impedances of power grid
$[Z_{Gnd,FC}]$	impedance of grounding connection between reference ground (Gnd) and FC
$[Z_{Gnd,M-FC}]$	impedance of grounding connection between motor and FC excluding cable shield
$[Z_{Sh,MC}]$	impedance of motor feeding cable shield

Common mode currents commonly encountered in ASD are circulating in two fundamental loops presented in Figure 1.4. From the perspective of FC, these two primary CM current loops are usually named as input side loop $I_{CM,In}$ and output side loop $I_{CM,Out}$. Both of CM currents depend on internal parameters of FC, which form a main coupling path between these two loops. The input side CM currents are additionally influenced by CM impedances of the power grid and grounding connection between FC and the power grid. The output-side CM currents are influenced by CM impedances of motor windings and feeding cable. Both of CM currents $I_{CM,In}$ and $I_{CM,Out}$ are always in part flowing through parasitic capacitive coupling that exists in FC, motor, and motor feeding cable that is indicated by dotted lines. Furthermore, two components of output CM current of the FC $I_{CM,Out}$ can be distinguished : $I_{CM,M}$—the most remote, flowing through motor windings, and $I_{CM,MC}$,— flowing in a smaller loop, through motor feeding cable only (Figure 1.4).

Based on the CMC distribution presented in Figures 1.3 and 1.4, it can be noticed that

- total CMC of FC is a sum of input CMC and output CMC (1.1) (Figure 1.3):

$$I_{CM,FC} = I_{CM,Out} + (-I_{CM,In}) \tag{1.1}$$

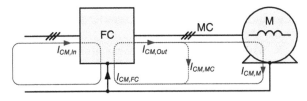

Figure 1.4 Fundamental CM currents loops in ASD.

- output CMC of FC is a sum of motor cable CMC (smaller loop) and motor CMC (larger loop) (1.2) (Figure 1.3):

$$I_{CM,Out} = I_{CM,MC} + I_{CM,M} \qquad (1.2)$$

- motor CMC can return back to FC through cable shield grounding connection and by other motor grounding connections that usually exist (1.3) (Figure 1.3),

$$I_{CM,M} = I_{CM,ShM} + I_{CM,Gnd} \qquad (1.3)$$

- output CMC of FC is also a sum of CMC returning back through motor cable shield and by other motor grounding connections that usually exist (1.4) (Figure 1.3).

$$I_{CM,Out} = I_{CM,Sh@FC} + I_{CM,Gnd} \qquad (1.4)$$

References

1 A. Anthon, Z. Zhang, and M. A. E. Andersen, "Comparison of a state of the art Si IGBT and next generation fast switching devices in a 4 kW boost converter," in *2015 IEEE Energy Conversion Congress and Exposition (ECCE)*, Sept. 2015, pp. 3003–3011.

2 X. Yang, Z. Long, Y. Wen, H. Huang, and P. R. Palmer, "Investigation of the trade-off between switching losses and EMI generation in Gaussian s-shaping for high-power IGBT switching transients by active voltage control," *IET Power Electronics*, vol. 9, no. 9, pp. 1979–1984, 2016.

3 N. Oswald, B. Stark, D. Holliday, C. Hargis, and B. Drury, "Analysis of shaped pulse transitions in power electronic switching waveforms for reduced EMI generation," *IEEE Transactions on Industry Applications*, vol. 47, no. 5, pp. 2154–2165, Sept.–Oct. 2011.

4 A. Srisawang, "A study of EMI and switching loss reductions of unipolar and improved limited unipolar switching circuits," in *2010 International Conference on Electrical Engineering/Electronics Computer Telecommunications and Information Technology (ECTI-CON)*, 2010, pp. 1211–1215.

5 J. Luszcz and K. Iwan, "Conducted EMI propagation in inverter-fed AC motor," *Electrical Power Quality and Utilisation, Magazine*, vol. 2, no. 1, 2006, pp. 47–51.

6 G. Skibinski, J. Pankau, R. Sladky, and J. Campbell, "Generation, control and regulation of EMI from AC drives," in *Industry Applications Conference, 1997. Thirty-Second IAS Annual Meeting, Conference Record of the 1997 IEEE*, vol. 2, 1997, pp. 1571–1583.

7 M. A. Chitsazan and A. M. Trzynadlowski, "Harmonic mitigation in interphase power controller using passive filter-based phase shifting transformer," in *2016 IEEE Energy Conversion Congress and Exposition (ECCE)*, Sept. 2016, pp. 1–5.

8 J. C. Das, *Passive Filters*. Wiley-IEEE Press, 2015. Available at http://ieeexplore. ieee.org/xpl/articleDetails.jsp?arnumber=7061327

9 T. Thasananutariya and S. Chatratana, "Planning study of harmonic filter for ASDs in industrial facilities," *IEEE Transactions on Industry Applications*, vol. 45, no. 1, pp. 295–302, Jan. 2009.

10 H. Bishnoi, P. Mattavelli, R. P. Burgos, and D. Boroyevich, "EMI filter design of DC-fed motor-drives using behavioral EMI models," in *2015 17th European Conference on Power Electronics and Applications (EPE'15 ECCE-Europe)*, Sept. 2015, pp. 1–10.

11 T. Nussbaumer, M. Heldwein, and J. Kolar, "Differential mode input filter design for a three-phase buck-type PWM rectifier based on modeling of the EMC test receiver," *IEEE Transactions on Industrial Electronics*, vol. 53, no. 5, pp. 1649–1661, 2006.

12 J. Kolar, U. Drofenik, J. Biela, M. Heldwein, H. Ertl, T. Friedli, and S. Round, "PWM converter power density barriers," in *Power Conversion Conference—Nagoya (PCC '07)*, 2007, pp. P-9–P-29.

13 A. Majid, J. Saleem, and K. Bertilsson, "EMI filter design for high frequency power converters," in *2012 11th International Conference on Environment and Electrical Engineering (EEEIC)*, 2012, pp. 586–589.

14 R. Wang, D. Boroyevich, H. Blanchette, and P. Mattavelli, "High power density EMI filter design with consideration of self-parasitic," in *2012 Twenty-Seventh Annual IEEE Applied Power Electronics Conference and Exposition (APEC)*, 2012, pp. 2285–2289.

15 J. Adabi, F. Zare, G. Ledwich, and A. Ghosh, "Leakage current and common mode voltage issues in modern AC drive systems," in *Power Engineering Conference, 2007. AUPEC 2007. Australasian Universities*, 2007, pp. 1–6.

16 Cichowlas, M. Malinowski, M. P. Kazmierkowski, D. L. Sobczuk, P. Rodriguez, and J. Pou, "Active filtering function of three-phase PWM boost rectifier under different line voltage conditions," *IEEE Transactions on Industry Electronics*, vol. 52, no. 2, pp. 410–419, 2005.

17 R. Smolenski, M. Jarnut, G. Benysek, and A. Kempski, "AC/DC/DC interfaces for v2g applications: EMC issues," *IEEE Transactions on Industrial Electronics*, vol. 60, no. 3, pp. 930–935, March 2013.

18 P. Nayak, M. V. Krishna, K. Vasudevakrishna, and K. Hatua, "Study of the effects of parasitic inductances and device capacitances on 1200 V, 35 a SiC MOSFET based voltage source inverter design," in *2014 IEEE International Conference on Power Electronics, Drives and Energy Systems (PEDES)*, Dec. 2014, pp. 1–6.

19 W. Sleszynski, J. Nieznanski, A. Cichowski, J. Luszcz, and A. Wojewodka, "Evaluation of selected diagnostic variables for the purpose of assessing the ageing effects in high-power igbts," in *2010 IEEE International Symposium on Industrial Electronics (ISIE)*, July 2010, pp. 821–825.

20 A. Kempski, R. Strzelecki, R. Smolenski, and G. Benysek, "Suppression of conducted EMI in four-quadrant AC drive system," in *2003 IEEE 34th Annual*

Power Electronics Specialist Conference (PESC'03), vol. 3, IEEE, 2003, pp. 1121–1126. Available at http://ieeexplore.ieee.org/xpls/abs_all.jsp?arnumber=1216606

21 J. Luszcz, "AC motor feeding cable consequences on EMC performance of ASD," in *2013 IEEE International Symposium on Electromagnetic Compatibility (EMC)*, 2013, pp. 248–252. Available at http://ieeexplore.ieee.org/stamp/stamp.jsp?arnumber=6670418

22 S. Amarir and K. Al-Haddad, "A modeling technique to analyze the impact of inverter supply voltage and cable length on industrial motor-drives," *IEEE Transactions on Power Electronics*, vol. 23, no. 2, pp. 753–762, March 2008.

23 S. Ogasawara and H. Akagi, "Analysis and reduction of EMI conducted by a PWM inverter-fed AC motor drive system having long power cables," in *2000 IEEE 31st Annual Power Electronics Specialists Conference (PESC '00)*, vol. 2, IEEE, 2000, pp. 928–933. Available at http://ieeexplore.ieee.org/xpls/abs_all.jsp?arnumber=879938

24 G. Skibinski, W. Maslowski, and J. Pankau, "Installation considerations for IGBT AC drives," in *IEEE 1997 Annual Textile, Fiber, and Film Industry Technical Conference*, 1997.

25 N. Hanigovszki, J. Landkildehus, G. Spiazzi, and F. Blaabjerg, "An EMC evaluation of the use of unshielded motor cables in AC adjustable speed drive applications," *IEEE Transactions on Power Electronics*, vol. 21, no. 1, pp. 273–281, 2006.

26 C. Jettanasen, "Influence of power shielded cable and ground on distribution of common mode currents flowing in variable-speed AC motor drive systems," in *2010 Asia-Pacific Symposium on Electromagnetic Compatibility (APEMC)*, April 2010, pp. 953–956.

27 F. Costa, C. Gautier, B. Revol, J. Genoulaz, and B. Demoulin, "Modeling of the near-field electromagnetic radiation of power cables in automotives or aeronautics," *IEEE Transactions on Power Electronics*, vol. 28, no. 99, p. 1, 2012.

28 S. A. Pignari and A. Orlandi, "Long-cable effects on conducted emissions levels," *IEEE Transactions on Electromagnetic Compatibility*, vol. 45, no. 1, pp. 43–54, 2003. Available at http://ieeexplore.ieee.org/xpls/abs_all.jsp?arnumber=1180392

29 J. Luszcz, "Modeling of common mode currents induced by motor cable in converter fed AC motor drives," in *2011 IEEE International Symposium on Electromagnetic Compatibility (EMC)*, IEEE, 2011, pp. 459–464. Available at http://ieeexplore.ieee.org/xpls/abs_all.jsp?arnumber=6038355

30 G. L. Skibinski, R. M. Tallam, M. Pande, R. J. Kerkman, and D. W. Schlegel, "System design of adjustable speed drives, Part 2: system simulation and AC line interactions," *IEEE Industry Applications Magazine*, vol. 18, no. 4, pp. 61–74, July 2012.

31 A. von Jouanne and P. N. Enjeti, "Design considerations for an inverter output filter to mitigate the effects of long motor leads in ASD applications," *IEEE Transactions on Industry Applications*, vol. 33, no. 5, pp. 1138–1145, Sept. 1997.

32 R. Strzelecki and G. Benysek, *Power Electronics in Smart Electrical Energy Networks.* Springer-Verlag, London, 2008.

33 R. Timens, F. Buesink, and F. Leferink, "Voltage quality in urban and rural areas," in *2012 IEEE International Symposium on Electromagnetic Compatibility (EMC)*, Aug. 2012, pp. 755–759.

34 Q. Yu and R. J. Johnson, "Smart grid communications equipment: EMI, safety, and environmental compliance testing considerations," *Bell Labs Technical Journal*, vol. 16, no. 3, pp. 109–131, Dec. 2011.

35 M. H. Bollen, J. Zhong, F. Zavoda, J. Meyer, A. McEachern, and F. Lopez, "Power quality aspects of smart grids," in *Proceedings of International Conference on Renewable Energies and Power Quality*, 2010.

36 R. Smolenski, A. Kempski, J. Bojarski, and P. Lezynski, "Zero cm voltage multilevel inverters for smart grid applications," in *2011 IEEE International Symposium on Electromagnetic Compatibility (EMC)*, 2011, pp. 448–453. Available at http://ieeexplore.ieee.org/stamp/stamp.jsp?arnumber=6038353

37 M. Olofsson, "Power quality and EMC in smart grid," in *10th International Conference on Electrical Power Quality and Utilisation. (EPQU '09)*, Sept. 2009, pp. 1–6.

38 I. Setiawan, C. Keyer, and F. Leferink, "Smarter concepts for future EMI standards," in *2017 Asia-Pacific International Symposium on Electromagnetic Compatibility (APEMC)*, June 2017, pp. 47–49.

39 P. Strauss, T. Degner, W. Heckmann, I. Wasiak, P. Gburczyk, Z. Hanzelka, N. Hatziargyriou, T. Romanos, E. Zountouridou, and A. Dimeas, "International white book on the grid integration of static converters," in *10th International Conference on Electrical Power Quality and Utilisation, 2009 (EPQU, 2009)*, Sept. 2009, pp. 1–6.

40 K. Armstrong, "Opportunities in the risk management of EMC," in *2011 IEEE International Symposium on Electromagnetic Compatibility (EMC)*. IEEE, 2011, pp. 988–993.

41 R. Smolenski, A. Kempski, J. Bojarski, and P. Lezynski, "EMI generated by power electronic interfaces in smart grids," in *2012 International Symposium on Electromagnetic Compatibility (EMC Europe)*, 2012, pp. 1–6. Available at http://ieeexplore.ieee.org/stamp/stamp.jsp?arnumber=6396771

42 H. Jo and K. J. Han, "Estimation of radiation patterns from the stator winding of AC motors using array model," in *2016 IEEE International Symposium on Electromagnetic Compatibility (EMC)*, July 2016, pp. 868–873.

43 R. Smolenski, J. Bojarski, A. Kempski, and J. Luszcz, "Aggregated conducted interferences generated by group of asynchronous drives with deterministic and random modulation," in *2012 Asia-Pacific Symposium on*

Electromagnetic Compatibility (APEMC), 2012, pp. 293–296. Available at http://ieeexplore.ieee.org/stamp/stamp.jsp?arnumber=6238025

44 D. Han, S. Li, Y. Wu, W. Choi, and B. Sarlioglu, "Comparative analysis on conducted CM EMI emission of motor drives: WBG versus Si devices," *IEEE Transactions on Industrial Electronics*, vol. 64, no. 10, pp. 8353–8363, Oct. 2017.

45 N. Oswald, P. Anthony, N. McNeill, and B. H. Stark, "An experimental investigation of the tradeoff between switching losses and EMI generation with hard-switched all-Si, Si-SiC, and all-SiC device combinations," *IEEE Transactions on Power Electronics*, vol. 29, no. 5, pp. 2393–2407, May 2014.

46 A. Kosonen, J. Ahola, and P. Silventoinen, "Measurements of HF current propagation to low voltage grid through frequency converter," in *2007 European Conference on Power Electronics and Applications*, Sept. 2007, pp. 1–10.

47 *IEC 61000-6-3 Electromagnetic Compatibility (EMC)—Part 6-3: Generic Standards—Emission Standard for Residential, Commercial and Light-Industrial Environments*, International Electrotechnical Commission Std.

48 *IEC 61000-6-4 Electromagnetic Compatibility (EMC)—Part 6-4: Generic Standards—Emission Standard for Industrial Environments*, International Electrotechnical Commission Std.

49 *IEC 61800-3:2017 Adjustable Speed Electrical Power Drive Systems—Part 3: EMC Requirements and Specific Test Methods*, International Electrotechnical Commission Std.

50 *IEEE Standard 519 (1992): Recommended Practices and Requirements for Harmonic Control in Electrical Power Systems*, International Electrotechnical Commission Std., 1992.

51 S. Chen, "Generation and suppression of conducted EMI from inverter-fed motor drives," in *Industry Applications Conference, 1999. Thirty-Fourth IAS Annual Meeting. Conference Record of the 1999 IEEE*, vol. 3, 1999, pp. 1583–1589.

52 M. Jin, M. WeiMing, P. Qijun, K. Jun, Z. Lei, and Z. Zhihua, "Identification of essential coupling path models for conducted EMI prediction in switching power converters," *IEEE Transactions on Power Electronics*, vol. 21, no. 6, pp. 1795–1803, Nov. 2006.

53 M. Hartmann, H. Ertl, and J. W. Kolar, "EMI filter design for a 1 MHz, 10 kW three-phase/level PWM rectifier," *IEEE Transactions on Power Electronics*, vol. 26, no. 4, pp. 1192–1204, April 2011.

54 J. Luszcz and R. Smolenski, "Low frequency conducted emissions of grid connected static converters," *IEEE Electromagnetic Compatibility Magazine*, vol. 4, no. 1, pp. 86–94, 2015.

55 S. G. Parker, D. S. Segaran, D. G. Holmes, and B. P. McGrath, "DC bus voltage EMI mitigation in three-phase active rectifiers using a virtual neutral filter," in *2014 International Power Electronics Conference (IPEC-Hiroshima 2014—ECCE ASIA)*, May 2014, pp. 2372–2379.

56 V. Xuning Zhang, D. Boroyevich, P. Mattavelli, J. Xue, and F. Wang, "EMI filter design and optimization for both AC and DC side in a DC-fed motor drive system," in *2013 Twenty-Eighth Annual IEEE Applied Power Electronics Conference and Exposition (APEC)*, 2013, pp. 597–603. Available at http://ieeexplore.ieee.org/stamp/stamp.jsp?arnumber=6520271

57 H. Akagi and T. Shimizu, "Attenuation of conducted EMI emissions from an inverter-driven motor," *IEEE Transactions on Power Electronics*, vol. 23, no. 1, pp. 282–290, 2008.

58 M. Dong, L. Zhai, R. Gao, and X. Zhang, "Research on radiated electromagnetic interference (EMI) from power cables of a three-phase inverter for electric vehicles," in *2014 IEEE Conference and Expo Transportation Electrification Asia-Pacific (ITEC Asia-Pacific)*, Aug. 2014, pp. 1–5.

59 M. Jin, Z. Lei, M. Weiming, Z. Zhihua, and P. Qijun, "Common-mode current inductively coupled emission of AC PWM drives," in *Asia-Pacific Symposium on Electromagnetic Compatibility and 19th International Zurich Symposium on Electromagnetic Compatibility, 2008. APEMC, PWM* 2008, pp. 650–653. Available at http://ieeexplore.ieee.org/stamp/stamp.jsp?arnumber=4559959

60 B. A. Acharya. and V. John, "Common mode DC bus filter for active front-end converter," in *2010 Joint International Conference on Power Electronics, Drives and Energy Systems 2010 Power India*, Dec. 2010, pp. 1–6.

61 L. Xing and J. Sun, "Conducted common-mode Emi reduction by impedance balancing," *IEEE Transactions on Power Electronics*, , vol. 27, no. 3, pp. 1084–1089, March 2012.

62 A. Kempski, R. Strzelecki, R. Smolenski, and Z. Fedyczak, "Bearing current path and pulse rate in PWM-inverter-fed induction," in *IEEE 32nd Annual Power Electronics Specialists Conference (PESC 2001)*, vol. 4, IEEE, 2001, pp. 2025–2030. Available at http://ieeexplore.ieee.org/xpls/abs_all.jsp?arnumber=954419

63 A. Muetze and A. Binder, "Generation of high-frequency common mode currents in machines of inverter-based drive systems," in *2005 European Conference on Power Electronics and Applications*, 2005. Available at http://ieeexplore.ieee.org/stamp/stamp.jsp?arnumber=1665679

64 A. von Jouanne, H. Zhang, and A. K. Wallace, "An evaluation of mitigation techniques for bearing currents, EMI and overvoltages in ASD applications," *IEEE Transactions on Industry Applications*, vol. 34, no. 5, pp. 1113–1122, 1998. Available at http://ieeexplore.ieee.org/xpls/abs_all.jsp?arnumber=720452

65 J. L. Kotny and N. Idir, "Time domain models of the EMI sources in the variable speed drives," in *2010 IEEE Energy Conversion Congress and Exposition (ECCE)*, 2010, pp. 1355–1360. Available at http://ieeexplore.ieee.org/stamp/stamp.jsp?arnumber=5618276

66 M. Moreau, N. Idir, and P. L. Moigne, "Modeling of conducted EMI in adjustable speed drives," *IEEE Transactions on Electromagnetic Compatibility*, vol. 51, no. 3, pp. 665–672, Aug. 2009.

67 C. Jettanasen, "Reduction of common-mode voltage generated by voltage-source inverter using proper PWM strategy," in *2012 Asia-Pacific Symposium on Electromagnetic Compatibility*, May 2012, pp. 297–300.

68 A. M. Trzynadlowski, K. Borisov, Y. Li, and L. Qin, "A novel random PWM technique with low computational overhead and constant sampling frequency for high-volume, low-cost applications," *IEEE Transactions on Power Electronics*, vol. 20, no. 1, pp. 116–122, Jan. 2005.

69 B. Singh, S. Gairola, B. Singh, A. Chandra, and K. Al-Haddad, "Multipulse AC–DC converters for improving power quality: a review," *IEEE Transactions on Power Electronics*, vol. 23, no. 1, pp. 260–281, 2008.

70 H.-D. Lee and S.-K. Sul, "A common-mode voltage reduction in boost rectifier/inverter system by shifting active voltage vector in a control period," *IEEE Transactions on Power Electronics*, vol. 15, no. 6, pp. 1094–1101, Nov. 2000.

71 A. Hota, S. Jain, and V. Agarwal, "A modified t-structured three-level inverter configuration optimized with respect to PWM strategy used for common mode voltage elimination," *IEEE Transactions on Industry Applications*, vol. 53, no. 99, pp. 1–1, 2017.

72 L. Wang, Y. Shi, and H. Li, "Anti-EMI noise digital filter design for a 60 kW 5-level SiC inverter without fiber isolation," *IEEE Transactions on Power Electronics*, vol. 33, no. 1, pp. 13–17, 2018.

73 H. Akagi and S. Tamura, "A passive EMI filter for eliminating both bearing current and ground leakage current from an inverter-driven motor," *IEEE Transactions on Power Electronics*, vol. 21, no. 5, pp. 1459–1469, 2006.

74 C. Jettanasen, "Passive common-mode EMI filter adapted to an adjustable-speed AC motor drive," in *2010 Conference Proceedings IPEC*, Oct. 2010, pp. 1025–1030.

75 R. K. Gupta, K. K. Mohapatra, A. Somani, and N. Mohan, "Direct-matrix-converter-based drive for a three-phase open-end-winding AC machine with advanced features," *IEEE Transactions on Industrial Electronics*, vol. 57, no. 12, pp. 4032–4042, Dec. 2010.

76 G. Skibinski, R. Tallam, R. Reese, B. Buchholz, and R. Lukaszewski, "Common mode and differential mode analysis of three phase cables for PWM AC drives," in *Industry Applications Conference, 2006. 41st IAS Annual Meeting. Conference Record of the 2006 IEEE*, vol. 2, 2006, pp. 880–888. Available at http://ieeexplore.ieee.org/stamp/stamp.jsp?arnumber=4025315

77 L. Zhou and S. A. Boggs, "High frequency attenuating cable for protection of low-voltage AC motors fed by PWM inverters," *IEEE Transactions on Power Delivery*, vol. 20, no. 2, pp. 548–553, April 2005.

78 X. Gong and J. A. Ferreira, "Comparison and reduction of conducted EMI in SiC JFET and Si IGBT-based motor drives," *IEEE Transactions on Power Electronics*, vol. 29, no. 4, pp. 1757–1767, April 2014.

2

Conducted Emission Origins in Switch-Mode Power Converters

If you want to find the secrets of the universe, think in terms of energy, frequency, and vibration.

Nikola Tesla

2.1 Spectral Side Effects of Fast Switching of Voltages and Currents

Switch-mode power conversion methods are based on fast switching of power electronics switches that allows controlling averaged values of voltages and currents easily with much higher efficiency than in case of using linear regulators. The efficiency achieved using those methods depends mainly on switching time duration. Generally, shorter switching time results in lower switching losses W_{sw} (2.1):

$$W_{sw} = \int_{t_0}^{t_0+t_{on}} v(t)i(t)dt \qquad (2.1)$$

Various methods are used to decrease switching losses for the given switching duration based on appropriate shaping of voltage and current waveforms during switching process. Nevertheless, all these solutions that allow decreasing switching losses usually result in increase of time derivatives of voltages (dv/dt) and currents (di/dt) [1–4]. From EMC point of view, any voltage and current changes arising in switch-mode power converters result in potential unintentional electromagnetic emission that has to be maintained within rational limits [5,6]. This interdependence imposes one of the most essential trade-offs in designing switch-mode power converters: between converter's

High Frequency Conducted Emission in AC Motor Drives Fed by Frequency Converters: Sources and Propagation Paths, First Edition. Jaroslaw Luszcz.
© 2018 by The Institute of Electrical and Electronic Engineers, Inc. Published 2018 by John Wiley & Sons, Inc.

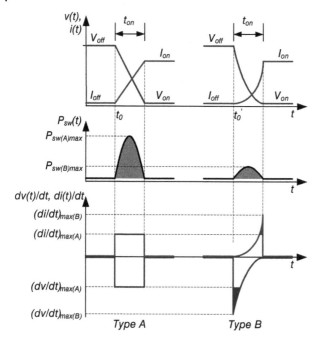

Figure 2.1 Transistor switching losses for different shapes of voltage and current waveforms.

general performance related to its efficiency and power density, and accompanied electromagnetic emission [7–9].

In Figure 2.1, two simplified waveforms of switched voltage and current are compared in terms of power losses and EMC consequences. One of them, marked as type A, represents linear voltage and current changes during switching used for approximated rough analysis. Second one, marked as B, represents exponential changes that are more realistic but still significantly simplified with reference to true transistor behavior. In case B, where during switching-on process voltage decreases faster and current increases slower than in case A, a significant decrease of total switching losses can be achieved (2.2), which is correlated with maximum switching power losses (2.3):

$$\int_{t_0}^{t_0+t_{on}} v(t)i(t)dt > \int_{t_0'}^{t_0'+t_{on}} v(t)i(t)dt \tag{2.2}$$

$$P_{SW(A)max} = \frac{1}{4}V_{off}I_{on} > P_{SW(B)max} \tag{2.3}$$

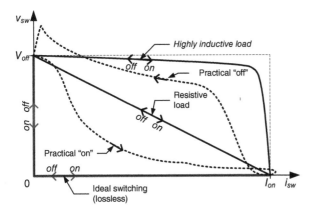

Figure 2.2 Typical simplified transistor's switching trajectories.

Generally, to minimize switching losses, the increase of switched-on current should be delayed until voltage decreases. It is possible to achieve within limited range by suitable adaptation of commutation circuit inductances and capacitances by using various techniques, for example, snubbers.

The theoretical classification of transistor switching trajectories is presented in Figure 2.2. According to this figure, switching trajectory for ideally resistive load can be used as reference with linear voltage and current changes (Figure 2.1). In practical applications, even if the load seems to be resistive, it usually includes some parasitic inductances and capacitances that change significantly its impedance in HF range. Therefore, presented practical trajectories for *on* and *off* transitions differ linearly and also between themselves due to parasitics. Appropriate modification of impedance characteristic of load and commutation circuits allows optimizing switching process to obtain the preferred objectives. Unfortunately, in practical applications, ideal lossless "zero" current switching is impossible to obtain fully.

Unfortunately, minimization of losses with the use of this technology results in the increase of total switching time t_{on} or consequently increases voltage and current derivatives, for the same value of switching time. Corresponding waveforms of derivatives are presented in Figure 2.1. Voltage and current waveform improvement in terms of switching losses results in increase of the values of current derivative of $(di/dt)_{max(A)}$ to $(di/dt)_{max(B)}$ for linear and exponential slopes, respectively. Similarly, for voltage waveforms, derivative of $(dv/dt)_{max(A)}$ for linear slope changes to $dv/dt_{max(B)}$ for exponential slope.

High (dv/dt) and (di/dt) are widely recognized as fundamental source of electromagnetic emission of switch-mode converters in various frequency bands. Any unintentional emission produced by switch-mode converters are commonly unwelcome and can be also harmful for other equipment or common electromagnetic environment, for example, power grids. The key issue is that more rapid changes of voltages and currents result in electromagnetic emission within broader frequency band. Limitation of EMI emission, which is required in many applications to ensure uninterrupted operation, usually becomes significantly more difficult and expensive to apply with the increase of frequency.

Spectrum of EM emission of switch-mode converters usually starts from low frequencies like power line frequency, causing side effects, for example, power system harmonics and audible noise. Nevertheless, the most challenging problems are observed in contemporary power electronics converters with fast switching devices, whose emission easily achieves frequency bands of conducted and radiated EMI, up to 30 MHz and above 30 MHz, respectively [10,11].

2.2 Spectra Estimation of Switch-Mode Waveforms

Effective determination of spectrum contents of voltage and current switching waveforms is very beneficial, because it allows predicting emission levels and, what is often even more important, evaluating influence of switching shapes on expected spectra contents. Improvement of switching waveform characteristic in order to obtain better EMC performance of switch-mode power conversion is the most effective solution, because it allows limiting EMI emission at very initial stage of design instead of attenuating EMI that are already generated and physically exist in some parts of a converter. Nowadays, a lot of research is focused on development of controlled switching of IGBT transistors to optimize their EMI emission end efficiency performance [12–14].

Unfortunately, the real switching waveforms of voltage and current changes in switching devices typically used in contemporary converters, for example, IGBT transistors, are complex and dependent on many parameters. Switching characteristic depends mostly on dynamic parameters of transistors itself, but also on parameters of external components like switched load, especially its inductance and capacitance [15,16]. Adequate selection of these parameters allows optimizing primary switching performance, like losses, and also other aspects, like accompanied EMI emission. Therefore, simplified analyses are widely used to estimate general relationship between switching waveform shapes and their spectral content because it requires only reasonable efforts to estimate essential parameters [17,18].

According to Fourier transform, any periodic waveform $x(t)$ represented in time domain can be represented as an equivalent harmonic content in frequency domain (2.4) where $X(k)$ are amplitudes of individual harmonics. The power of the transmitted signal in time domain has to be equal to the sum of powers of all harmonic components, according to Parseval's theorem. This rule defines power density spectrum of signals as power associated with all

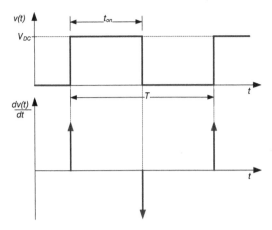

Figure 2.3 Parameters of rectangular waveform.

frequency components of the periodic signal (2.4). The power of periodic signal is equal to the sum of the powers of its harmonic components:

$$\frac{1}{T}\int_{t_0}^{t_0+T}|f(t)|^2\,dt = \sum_{k=-\infty}^{\infty}|X(k)|^2 \tag{2.4}$$

2.2.1 Rectangular Waveforms

Fourier transform of a periodic rectangular waveform allows determining magnitudes c_k and angle coefficients θ_k of its spectrum defined as (2.5) [19]:

$$x(t) = c_0 + \sum_{k=1}^{\infty} c_k \cos\left(k\frac{2\pi}{T}t + \theta_k\right) \tag{2.5}$$

$$\text{where } c_k = |X(k)| \quad \text{and} \quad \theta_k = \angle X(k)$$

Considering specific properties of rectangular waveform and taking into account the properties of Fourier transform, a formula for obtaining coefficients c_n of one side spectrum can be derived [19]. It means that the envelope of amplitudes of spectral components c_n of rectangular wave can be formulated as sinc function (2.6), where $f_1 = 1/T$:

$$|c_n| = V_{DC}\frac{t_{on}}{T}\left|\frac{\sin(n\pi t_{on}f_1)}{n\pi t_{on}f_1}\right| \tag{2.6}$$

Analytic calculation of Fourier transform, even for simplified switch-mode waveforms like the one of rectangular shape, is rather complex (Figure 2.3).

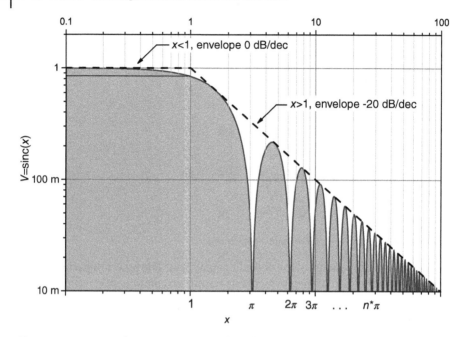

Figure 2.4 Linear envelope approximation of sinc function.

Therefore, simplified methods to estimate waveforms spectra based on sinc function have been in use for a long time [19,20].

The properties of sinc function (2.6) significantly simplify determination of maximum values of expected spectrum called "envelope," which is most essential for EMC analysis. The properties of sinc function (2.7) clearly represent the properties of emission spectrum of single rectangular pulse that are presented in Figure 2.4.

$$|\text{sinc}(x)| = \left| \frac{\sin(x)}{x} \right| \leq \begin{cases} 1, & \text{for } x \ll 1 \\ \frac{1}{|x|}, & \text{for } x \gg 1 \end{cases} \tag{2.7}$$

Spectrum of a waveform that can be approximated as periodically repeated rectangular pulse train with repetition period T can also be estimated using sinc(x) function. The envelope boundary, which determines maximum values of harmonic magnitudes, can be characterized in frequency domain assuming that $x = \pi t_{on}/T$. However, spectrum of rectangular pulse train has noncontinuous but line character where each spectrum line is located at multiple of pulse repetition frequency $f_1 = 1/T$ (Figure 2.5).

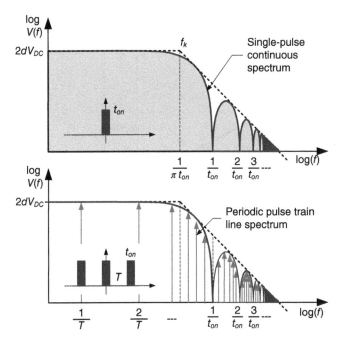

Figure 2.5 Continuous spectrum envelope of single rectangular pulse (upper characteristic) and line spectrum of rectangular pulse train.

Considering the properties of sinc(x) function 2.1, it can be noticed that for rectangular waveform with repetition frequency equal to $f_1 = 1/T$:

- magnitudes of spectrum components generally follow the form of function $sinc(x)/x$, although it appears only for frequencies $f_n = n/T$,
- magnitudes of spectrum components go to zero for multiples of π $x = k\pi$, which means that zero amplitudes of spectrum will occur for frequencies equal to multiples of $f_{on} = n/t_{on}$,
- magnitudes of spectrum components tend to unity for $x < 1$, in frequency domain for $f < 1/t_{on}$
- for $x > 1$, $f < 1/t_{on}$ exhibits characteristic bounds between zero values occurring for $f_{on} = n/t_{on}$. Maximum values of these bounds can be enveloped by the asymptote $1/x$, which is sloped -20 dB per decade.

Characteristic shape of normalized spectrum envelope calculated according to the formula (2.6) is presented in Figure 2.6. Assuming reference level 0 dB as the value of switched DC voltage, the beginning of spectrum envelope for pulse

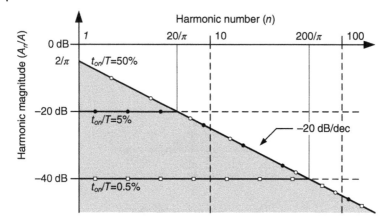

Figure 2.6 Normalized envelope of rectangular waveform with different duty cycles.

train with duty cycle equal to 50% starts from the value $2/\pi \approx -3.9$ dB for normalized frequency $f = 1$ and decreases with frequency with slope -20 dB/dec. For duty cycle lower than 50%, the characteristic knee frequency can be determined as $f_k = 1/(\pi\delta)$. Thus, for $\delta = 5\%$ and 0.5%, we have characteristic frequencies $f_k = 20/(\pi\delta)$ and $200/(\pi\delta)$, respectively. For rectangular pulse trains with duty cycle smaller than 50%, spectrum envelope is flat up to the knee frequency and equals $2 * \delta$; therefore, for $\delta = 5\%$ and 0.5%, we have -20 and -40 dB, respectively. Finally, with the decrease of duty cycle of the rectangular waveform, the flat part of spectrum envelope extends toward higher frequencies and its magnitude in this frequency range decreases proportionally to δ.

Summarizing the above consideration, the influence of duty cycle of rectangular waveform defined as $\delta = t_{on}/T$ can be analyzed based on formula (2.6). For rectangular waveform with duty cycle $\delta = 50\%$, which is the maximum, the spectrum envelope decreases with frequency with -20 dB per decade slope. For duty cycle δ lower than 50%, the spectrum envelope tends to constant value equal to $2\delta t_{on}/T$ up to the knee frequency $f_k = 1/\pi t_{on}$ and above this frequency is sloping down -20 dB per decade. Exemplary results obtained for rectangular waveform, which is a simplified representation of waveforms typically occurring in ASD, are presented in Figure 2.7.

Since the parameters of the analyzed waveform are close to those really appearing in typical power converters, based on this graphical approximation the expected maximal magnitudes of spectrum versus frequency can be estimated theoretically. Magnitudes of generated harmonics resulting from such switching can achieve levels of few volts in frequency range around 1 MHz and

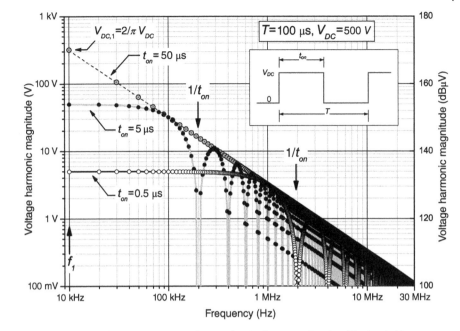

Figure 2.7 Exemplary spectrum envelopes of typical rectangular simplified switching waveform, occurring in power electronic converter: pulse repetition frequency $f_1 = 10$ kHz, switched voltage $V_{DC} = 500$ V and different duty cycles $\delta = 50, 5,$ and 0.5% obtained for different switch-on times 50, 5, and 0.5 μs, respectively.

decrease to few hundreds of millivolts for frequencies close to the upper limit of conducted emission, that is, 30 MHz.

2.2.2 Trapezoidal Waveforms

Trapezoidal approximations of switching waveforms appearing in switched-mode power electronic converters are used for more accurate analysis. This is because they allow taking into account switch-on and switch-off times that are essential in spectrum analysis. Analysis based on trapezoidal approximation allows obtaining relatively valuable results. Nevertheless, adequacy of this type of waveform approximation can be problematic in many applications. General parameters of trapezoidal approximation are defined in Figure 2.8 similarly as for the rectangular waveform (Figure 2.3) with two additional parameters: rise time t_r and fall time t_f.

For a trapezoidal waveform, two derivatives can be determined: first one that is rectangular function with pulse duration related to voltage slopes, and the second one that is impulse function appears at each beginning and end of rise and fall slopes [19]. According to Figure 2.9, a simplified trapezoidal

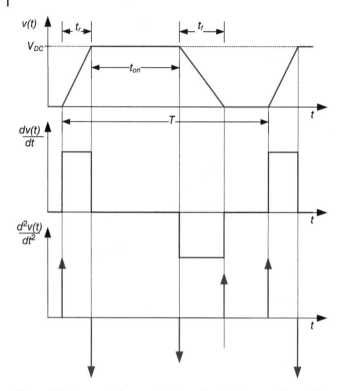

Figure 2.8 Trapezoidal approximation of switching waveforms: V_{DC}—switched DC voltage; T—switching period; t_{on}—voltage pulse duration; t_r—voltage rise time; t_f—voltage fall time.

waveform $f_{tr}(t)$ with equal rise and fall times ($t_r = t_f$) can be considered as the convolution product of ideally rectangular waveform $f_{sq}(t)$ with another rectangular function that is a normalized derivative $f_{nd}(t)$ of the analyzed trapezoidal waveform $f_{tr}(t)$. Parameters of convoluted functions that produce a trapezoidal waveform are presented in more detail in Figure 2.9.

According to convolution properties, spectrum content of trapezoidal waveform represented as convolution product of two rectangular waveforms in time domain (2.8) can be considered as product of individual spectra of both rectangular waveforms in frequency domain (2.9):

$$f_{tr}(t) = f_{sq}(t) * f_{nd}(t) \tag{2.8}$$

$$f_{tr}(f) = f_{sq}(f) \times f_{nd}(f) \tag{2.9}$$

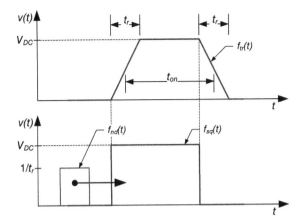

Figure 2.9 Trapezoidal waveform represented as a convolution of two ideally rectangular waveforms.

Finally, taking into account the above consideration and the formula (2.6) for spectrum estimation of a rectangular waveform presented in 2.2.1, the corresponding formula for estimation of trapezoidal waveform spectra envelope can be formulated as (2.10).

$$|c_n| = 2V_{DC}\frac{t_{on}}{T} \left| \frac{\sin(n\pi t_{on}f_1)}{n\pi t_{on}f_1} \right| \left| \frac{\sin(n\pi t_r f_1)}{n\pi t_r f_1} \right| \tag{2.10}$$

Graphical representation of spectrum envelope of trapezoidal waveform with two characteristic knee frequencies f_{k1} and f_{k2} is presented in Figure 2.10. In the frequency range above the first characteristic knee frequency f_{k1}, which is related to switch-on time $f_{k1} = 1/(\pi t_{on})$, the amplitudes of spectrum components are decreasing with the slope −20 dB/dec similar to the rectangular waveform. Above the second characteristic knee frequency f_{k2} that is related to rise time $f_{k2} = 1/(\pi t_r)$ the magnitudes of spectrum components are decreasing with the slope −40 dB/dec, which is the effect of rectangular shape of first derivative of trapezoidal waveform.

Exemplary results of detailed calculation of spectrum envelope using formula (2.10) are presented in Figure 2.11. In the analyzed waveform, PWM carrier frequency f_1 = 10 kHz and switched voltage V_{DC} = 500 V are the same as in previous analysis for rectangular waveform (Figure 2.7). However, additional parameter, switching time $t_r = t_f$ = 100 ns, is introduced. Taking switching times into account allows estimating generated spectrum more adequately in the frequency range above the second knee $f_{k2} = 1/(\pi t_r)$ where spectrum magnitudes are decreasing −40 dB/dec.

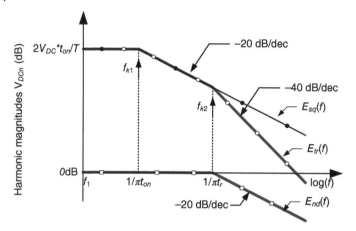

Figure 2.10 Simplified linear spectrum envelope of trapezoidal waveform with rise time equal to fall time.

In power electronic applications where usual power transistors are used, switching-on and switching-off times are usually significantly different in length and should be approximated by trapezoidal waveforms with different durations

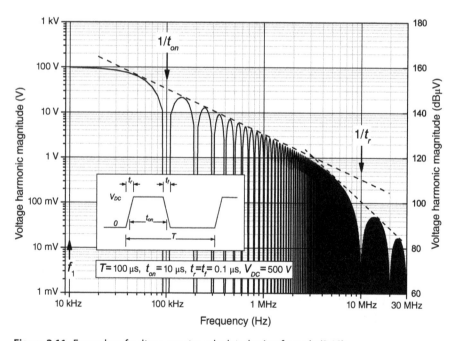

Figure 2.11 Examples of voltage spectra calculated using formula (2.10).

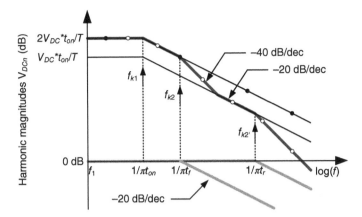

Figure 2.12 Analysis of spectrum envelop of trapezoidal waveform with different rise time and fall time.

of rising and falling slopes. Therefore, analysis of such cases is in fact more beneficial, but unfortunately becomes more complex. As an example, detailed analysis of resultant spectra for trapezoidal waveform with two different slopes is presented in Figure 2.12.

Effects of −40 dB/dec spectrum envelope sloping as a result of different rise time and fall time start to take place for different frequencies $f_r = 1/\pi t_r$ and $f_f = 1/\pi t_f$. In between these two frequencies, spectrum envelope decreases −40 dB/dec until it obtains the level related to spectrum accompanying to rise time f_r and then starts sloping again but −20 dB/dec. Above the frequency f_r related to the rise time, the spectrum envelope starts decreasing again at −40 dB/dec. Based on this analysis, it can be noticed that for waveforms with many slopes with different durations, the most significant influence on the waveform spectrum envelope is the shortest slope time that determines the frequency above which the envelope starts to decrease −40 dB/dec. The remaining longer slopes included in waveform approximation are less significant and change waveform spectrum only slightly.

Development of graphical methods of analysis of trapezoidal waveforms with different slopes is beneficial, because it can be used in analysis of more complex waveforms by approximating them by a series of trapezoidal segments with different slopes $t_{r1} \cdot t_{rN}$. Using this methodology, waveforms generated by real-power transistors can be modeled more accurately (Figure 2.11). The spectrum envelope of such a waveform can be considered as a total effect of spectra associated with each sloping part of the waveform. Nevertheless, it can be noticed that the final waveform spectrum envelope is mostly influenced by the slope segment with the highest voltage change in the shortest time, that meaning with the highest value of the first derivative. The impact of the

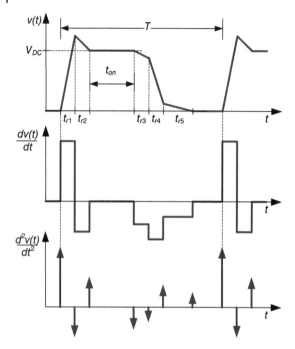

Figure 2.13 Transistor switching waveform approximation by a number of trapezoidal segments.

remaining slope segments with lower derivative is less meaningful for the final spectrum envelope.

Generally, trapezoidal approximation of waveforms can be described as waveforms with rectangular first derivatives, which results in −40 dB/dec decrease of spectrum possible to obtain (Figure 2.13). The graphical method of waveform spectrum estimation presented in this chapter is based on the assumption that spectrum envelope of single pulse is limited by two inequalities [21–23].

$$|F(\omega)| \leq \int_{-\infty}^{\infty} |f(t)|\, dt \tag{2.11}$$

$$|F(\omega)| \leq \frac{\int_{-\infty}^{\infty} \left| \frac{d^n f(t)}{dt^n} \right| dt}{\omega^n} \tag{2.12}$$

Based on Eq. (2.12), it can be noticed that for graphical analysis a waveform should be derived until the first appearance of impulse function [19]. Furthermore, decrease of harmonic amplitudes in high-frequency range is correlated with the number of derivatives of the analyzed waveforms. The highest order of derivative possible to obtain, d^n, allows estimating harmonic spectrum decrease above the characteristic knee frequency as $(n + 1) * (-20\,\text{dB/dec})$.

According to this rule for the trapezoidal shape, which can be derived only once (second derivative is already impulse function), it is possible to have maximally only $-40\,\text{dB/dec}$ spectrum decrease with frequency. Based on the above consideration, it can be concluded that the effective method to limit harmonic emission in high-frequency range can be based on special shaping of a switching waveform that are differentiable more than once.

2.2.3 Higher Order Transition Shapes

There are many studies focused on improvement of the transistor switching process toward decreasing accompanied harmonic emission [18,24]. It is possible to do by using special shaping of the transistor gate current to obtain smoother voltage switching. The considerations presented in the previous chapter help determine desired waveform shapes that allow limiting EMI emission efficiently. In practice applications, transistor switching waveforms can be changed only within a very limited range. Analysis of S-shaped wave, as a theoretical example of switching waveform with more than two derivatives, is presented in this chapter. An example of analyzed S-shaped waveform is presented in Figure 2.14. In this idealized case, the first derivative $dv(t)/dt$ of S-shaped waveform with equal rise and fall times (t_r, t_f) becomes trapezoidal with rise and fall times (t'_r, t'_f) that are also equal. The second derivative $dv^2(t)/dt$ has rectangular shape with pulse duration equal to t'_r and t'_f. The last, derivative $dv^3(t)/dt$ becomes impulse function.

According to the analysis of trapezoidal waveforms presented in previous Section 2.2.2, for the waveform with existing two derivatives the resulting spectrum can be estimated as a product of two convolution operations of rectangular waveform spectra correlated with pulse duration t_{on}, pulse rise time t_r, and rise time of first derivative t'_r. As for trapezoidal waveform that can be represented as convolution product of two rectangular waveforms in time domain (2.9), S-shaped waveform can be represented as a convolution product of trapezoidal waveform dv/dt and rectangular waveform with duty cycle $\delta = t_{on}/T$. The first derivative of waveform is already a convolution product of rectangular waveform d^2v/dt and rectangular waveform with duty cycle $\delta = t_r/T$. Based on the above assumption, the spectrum envelope of S-shaped waveform can be calculated according to Eq. (2.13).

$$|c_n| = 2V_{DC}\frac{t_{on}}{T}\left|\frac{\sin(n\pi t_{on}f_1)}{n\pi t_{on}f_1}\right|\left|\frac{\sin(n\pi t_r f_1)}{n\pi t_r f_1}\right|\left|\frac{\sin(n\pi t'_r f_1)}{n\pi t'_r f_1}\right| \qquad (2.13)$$

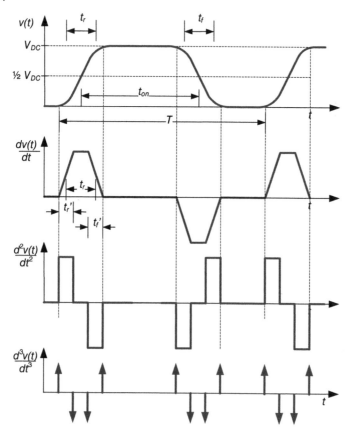

Figure 2.14 S-shaped voltage waveform as an example of higher order voltage transition.

The graphical representation of spectra of S-shaped waveform is presented in Figure 2.15. The presented envelope exhibits three knee frequencies correlated with pulse duration t_{on}, pulse rise time t_r, and rise time of pulse waveform first derivative t'_r. Based on this simplified graphical analysis of spectrum envelope, it can be noticed that by smoothing trapezoidal waveform vertices by S-shaped function its spectral content can be significantly limited, in frequency range above $1/(\pi t'_r)$, by obtaining −60 dB/dec slope of spectral characteristics. This relationship allows optimizing transistor switching process for decreasing its emission characteristics [25–27].

2.3 Graphical Estimation of Emission Spectra

The accuracy of graphical method of spectrum estimation can be verified by comparison with the results obtained using analytical methods based on FFT

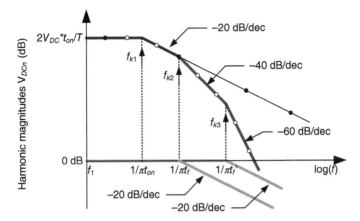

Figure 2.15 Simplified linearized spectra envelope of the S-shaped waveform.

analysis. An example of real switching voltage waveform typically appearing at three-phase transistor bridges supplied by DC voltage resulting from rectification of LV grid is presented in Figure 2.16. Closer view of transistor switch-off and switch-on processes shows significant differences between their shapes. Switch-off time is much longer than switch-on time. Both of the slopes are considerably different from trapezoidal shape: Switch-on shape can be classified as exponential, whereas switch-off shape has few trapezoidal steps of different time durations. Nevertheless, approximate switch-on and switch-off times can be graphically estimated as about 3.5 μs and 200 ns, respectively (Figure 2.16).

Assuming that the analyzed waveform is trapezoidal with different slopes and according to graphical methods presented in previous chapter, characteristic spectrum knee frequencies can be calculated: the first characteristic knee frequency $f_{k1} \approx 90$ kHz (2.14), and the second $f_{k2} \approx 1.6$ MHz (2.15):

$$f_{k1} = \frac{1}{\pi t_r} = \frac{1}{\pi 3.5 \mu s} \approx 90 \, \text{kHz} \tag{2.14}$$

$$f_{k1} = \frac{1}{\pi t_f} = \frac{1}{\pi 0.2 \mu s} \approx 1.6 \, \text{MHz} \tag{2.15}$$

Adequate linear asymptotes −20 and −40 dB/dec linked with the knee frequencies are compared in Figure 2.17 with the spectrum calculated analytically using FFT analysis. Determined characteristic knee frequencies can be recognized as analytical results and their correlation with graphically positioned asymptotes is noticeable.

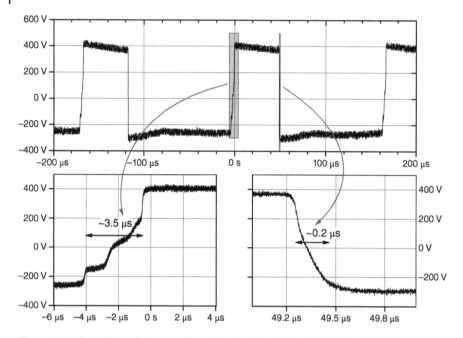

Figure 2.16 Switching voltage waveform at output terminals of full-bridge DC–AC inverter supplied by DC bus voltage originated from grid-powered diode rectifier.

2.4 Spectral Analysis of Switch-Mode Converters within Frequency Range of Conducted Emission

Graphical methods of spectra estimation in power electronic applications based on switching waveforms analysis can be very useful because they allow roughly approximate expected levels of emission with relatively little computational overheads. Nevertheless, in real applications, there are a number of additional physical phenomena that can change switching waveform shapes and thus the resultant emission. Therefore, theoretical analysis based on simplified graphical estimation method can be considered as very efficient first approximation, which require further verification. Contrary to graphical methods, more detailed analysis can allow obtaining much more accurate results, although the computational effort very often becomes enormously greater.

The most significant reasons that limit the adequacy of theoretical evaluations are related to variable duty cycle of voltage and current waveforms as a general rule of switching mode regulation. Furthermore, transistors switching characteristic depends on values of the switched current. This leads to significant differences in time domain of harmonic emission of typical converter; so some averaging methods are necessary to use. This averaging problem is

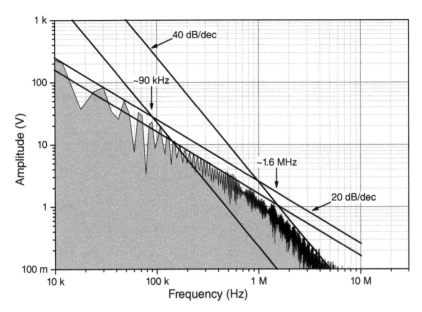

Figure 2.17 Spectrum of experimentally obtained voltage waveform presented in Figure 2.16 calculated using FFT method with graphically estimated linear asymptotes.

solved by implementation of standardized detectors recommended to use in EMI measurement systems.

Commonly, peak, quasi-peak, and average detectors are used, with standardized time constant that allows standardized measurement of emission of sources with periodically variable activity. Thus, in typical switch-mode converter, we usually have a number of line spectrum sources with slightly movable spectrum lines, intentionally or not. Rapid variations of generated harmonic spectra in time are difficult to analyze accurately; therefore, the specified measurement resolution bandwidths (RBW) of spectrum analyzer are necessary to use to avoid the measurement results discrepancy. Typical EMI measurements are usually carried out using standardized RBW specified by EMC standards [28,29].

Further problems, which can significantly change spectra estimated based on analysis of transistor switching waveforms using graphical methods, are related to voltage and current harmonics propagation through different electrical components used in power electronics converters, for example, inductors, capacitors, loads, and even wire connections. Broadband behavior of these devices can influence transmitted waveform shapes significantly, which also results in associated changes of harmonic emission spectra. An example of transistor switching waveform influenced by external load components is presented in Figure 2.18. This waveform has been obtained for the same converter output as

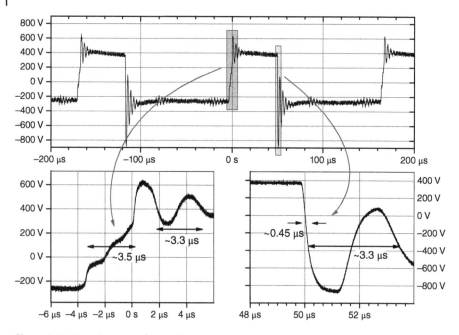

Figure 2.18 Transistor switching voltage waveform with ringing effects.

in Figure 2.16, although wideband characteristic of converter's load has been changed.

Changed broadband characteristic of load introduce clearly visible extra ringing effects and increase transistor switch-on time noticeably. These ringing effects are the results of load circuit inductances and capacitances and change harmonic emission spectrum selectively within frequency ranges correlated with oscillations frequency. In the presented example, frequency of ringing $f_r \approx 300$ kHz is clearly visible in the spectrum shape. Higher order harmonic components of ringing frequency are also noticeable, especially the third harmonic $f_{r3} \approx 1$ MHz. Spectral characteristic of this waveform (Figure 2.18) is presented in Figure 2.19 with the same asymptotes as in Figure 2.17. This comparison illustrates how ringing effects originating from other components than switching transistor can change final harmonic emission [30,31].

Accurate modeling of harmonic emission in switch-mode converters requires taking into account properties of all components of the whole power-conversion system, including load and supply network. Identification of parameters of all converter components in the frequency range adequate to expected emission is a key difficulty limiting effectiveness of this analysis. Broadband characteristics of typical components used in power conversion are very often nonlinear and difficult to determine accurately in a wide frequency range. Furthermore,

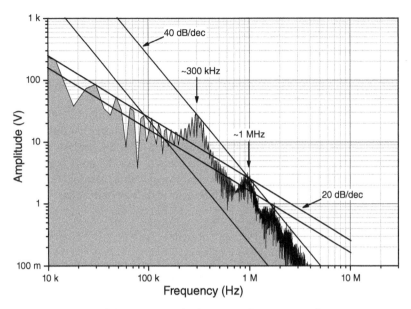

Figure 2.19 Spectral characteristic of voltage waveform presented in Figure 2.18.

with the increase of frequency parasitic couplings that exist inside components and also in-between components, not only electrical, become very significant. Effects of parasitic couplings can change considerably high-frequency harmonics propagation and cause leakages that are difficult to predict in theoretical analysis. Complete broadband analysis of EMI emission requires taking into consideration enormous number of parameters of the evaluated converter related not only to electrical components but also to mechanical, thermal and so on, and therefore usually becomes very complex and difficult to use efficiently. Furthermore, even external components of the power conversion system can influence its internal broadband behavior, because of possible resonant interactions [32].

2.5 Summary

The fundamental source of harmonic emission in switch-mode power converters are rapid voltage changes that appear during transistor switching. Analysis of transient waveforms at switching transistor terminals allows estimating expected harmonic emission levels. This analysis can be done either using simplified graphical methods presented in this chapter or analytically, more accurately, but with much grater computational efforts.

Final harmonic emission of switch-mode power converters also significantly depends on wideband characteristics of other components besides transistors; therefore, in many cases simplified graphical analysis can be satisfactory enough at preliminary study stage. Efficient simulation tools for global emission analysis of entire switch-mode power conversion system are very much required and developed intensively by many research groups.

There are already many well-developed useful tools for theoretical analysis of harmonic emission; nevertheless, their use does not guarantee full success in all applications. Despite availability of excellent tools for analysis of many particular issues of EMC in selected components of switch-mode converters, the main issue is actually related to accurate wideband identification of all components of the analyzed system in different working conditions.

References

1 D. Aggeler, F. Canales, J. Biela, and J. W. Kolar, "d v/d t-control methods for the SiC JFET/Si MOSFET cascode," *IEEE Transactions on Power Electronics*, vol. 28, no. 8, pp. 4074–4082, Aug. 2013.

2 N. Idir, R. Bausiere, and J. J. Franchaud, "Active gate voltage control of turn-on di/dt and turn-off dv/dt in insulated gate transistors," *IEEE Transactions on Power Electronics*, vol. 21, no. 4, pp. 849–855, 2006. Available at http://ieeexplore.ieee.org/xpls/abs_all.jsp?arnumber=1650284

3 J. Kagerbauer and T. Jahns, "Development of an active dv/dt control algorithm for reducing inverter conducted EMI with minimal impact on switching losses," in *Power Electronics Specialists Conference, 2007 (PESC'07), IEEE*, June 2007, pp. 894–900.

4 M. M. Swamy, J. K. Kang, and K. Shirabe, "Power loss, system efficiency, and leakage current comparison between Si IGBT VFD and SiC FET VFD with various filtering options," *IEEE Transactions on Industry Applications*, vol. 51, no. 5, pp. 3858–3866, Sept. 2015.

5 R. Monteiro, B. Borges, and V. Anunciada, "EMI reduction by optimizing the output voltage rise time and fail time in high-frequency soft-switching converters," in *2004 IEEE 35th Annual Power Electronics Specialists Conference, 2004 (PESC'04)*, vol. 2, 2004, pp. 1127–1132.

6 H. Soltani, F. Zare, and J. Adabi, "Effects of switching time on output voltages of a multilevel inverter used in high frequency applications," in *Power Engineering Conference, 2007 (AUPEC'07). Australasian Universities*, 2007, pp. 1–6.

7 K. Raggl, T. Nussbaumer, and J. W. Kolar, "Guideline for a simplified differential-mode EMI filter design," *IEEE Transactions on Industrial Electronics*, vol. 57, no. 3, pp. 1031–1040, March 2010.

8 R. Smolenski, "Selected conducted electromagnetic interference issues in distributed power systems," *Bulletin of the Polish Academy of Sciences: Technical Sciences*, vol. 57, no. 4, pp. 383–393, 2009.

9 M. R. Yazdani, H. Farzanehfard, and J. Faiz, "Classification and comparison of EMI mitigation techniques in switching power converters: a review," *Journal of Power Electronics*, vol. 11, no. 5, pp. 767–777, Sept. 2011.

10 J. Luszcz, "High frequency harmonics emission of modern power electronic AC–DC converters," in *8th International Conference on Compatibility and Power Electronics (CPE)*, 2013, pp. 269–274. Available at http://ieeexplore.ieee.org/stamp/stamp.jsp?arnumber=6601168

11 G. Skibinski, R. Kerkman, and D. Schlegel, "EMI emissions of modern PWM ac drives," *IEEE Industry Applications Magazine*, vol. 5, no. 6, pp. 47–80, 1999.

12 T. Igarashi, H. Funato, S. Ogasawara, M. Hara, and Y. Hirota, "Performance of power converter applied switching transient waveform modification," in *2010 International Power Electronics Conference (IPEC)*, June 2010, pp. 1882–1887.

13 A. Karvonen, H. Holst, T. Tuveson, T. Thiringer, and P. Futane, "Reduction of EMI in switched mode converters by shaped pulse transitions," in *SAE World Congress 2007*, April 16–19, 2007, Cobo Center, Detroit, MI, 2007.

14 J. Meng, W. Ma, Q. Pan, L. Zhang, and Z. Zhao, "Multiple slope switching waveform approximation to improve conducted EMI spectral analysis of power converters," *IEEE Transactions on Electromagnetic Compatibility*, vol. 48, no. 4, pp. 742–751, Nov. 2006.

15 N. Oswald, P. Anthony, N. McNeill, and B. H. Stark, "An experimental investigation of the tradeoff between switching losses and EMI generation with hard-switched all-Si, Si-SiC, and all-SiC device combinations," *IEEE Transactions on Power Electronics*, vol. 29, no. 5, pp. 2393–2407, May 2014.

16 G. Schmitt, R. Kennel, and J. Holtz, "Voltage gradient limitation of IGBTs by optimised gate-current profiles," in *Power Electronics Specialists Conference, 2008 (PESC'08), IEEE*, June 2008, pp. 3592–3596.

17 F. Mihalic and M. Milanovic, "Power density spectrum estimation of the random controlled PWM single-phase boost rectifier," in *1999 IEEE International Symposium on Electromagnetic Compatibility*, vol. 2, 1999, pp. 803–805. Available at http://ieeexplore.ieee.org/stamp/stamp.jsp?arnumber=810122

18 N. Oswald, B. Stark, D. Holliday, C. Hargis, and B. Drury, "Analysis of shaped pulse transitions in power electronic switching waveforms for reduced EMI generation," *IEEE Transactions on Industry Applications*, vol. 47, no. 5, pp. 2154–2165, Sept.–Oct. 2011.

19 C. R. Paul, *Introduction to Electromagnetic Compatibility*, John Wiley & Sons, Inc., 2006.

20 S. Ogasawara, T. Igarashi, H. Funato, and M. Hara, "Optimization of switching transient waveform to reduce EMI noise in a selective frequency band," in *Energy Conversion Congress and Exposition, 2009 (ECCE'09), IEEE*, Sept. 2009, pp. 1679–1684.

21 B. Audone, "Graphical harmonic analysis," *IEEE Transactions on Electromagnetic Compatibility*, vol. EMC-15, no. 2, pp. 72–74, May 1973.

22 A. W. Di Marzio, "Graphical solutions to harmonic analysis," *IEEE Transactions on Aerospace and Electronic Systems*, vol. AES-4, no. 5, pp. 693–706, Sept. 1968.

23 J. Hu, J. Von Bloh, and R. De Doncker, "Typical impulses in power electronics and their EMI characteristics," in *2004 IEEE 35th Annual Power Electronics Specialists Conference(PESC '04).*, vol. 4, 2004, pp. 3021–3027.

24 X. Yang, X. Zhang, and P. R. Palmer, "IGBT converters conducted EMI analysis by controlled multiple-slope switching waveform approximation," in *2013 IEEE International Symposium on Industrial Electronics*, May 2013, pp. 1–6.

25 J. J. O. Dalton, J. Wang, H. C. P. Dymond, D. Liu, D. Pamunuwa, B. H. Stark, N. McNeill, and S. J. Hollis, "Shaping switching waveforms in a 650 V GaN FET bridge-leg using 6.7 GHz active gate drivers," in *2017 IEEE Applied Power Electronics Conference and Exposition (APEC)*, March 2017, pp. 1983–1989.

26 N. F. Oswald, B. H. Stark, C. Hargis, W. Drury, and D. Holliday, "Synthesizing power electronic switching waveforms for reduced EMI generation," in *5th IET International Conference on Power Electronics, Machines and Drives (PEMD'10)*, April 2010, pp. 1–6.

27 X. Yang, Y. Yuan, X. Zhang, and P. R. Palmer, "Shaping high-power IGBT switching transitions by active voltage control for reduced EMI generation," *IEEE Transactions on Industry Applications*, vol. 51, no. 2, pp. 1669–1677, March 2015.

28 W. Schaefer, "Measurement of impulsive signals with a spectrum analyzer or EMI receiver," in *2005 International Symposium on Electromagnetic Compatibility, 2005 (EMC'05)*, vol. 1, 2005, pp. 267–271. Available at http://ieeexplore.ieee.org/stamp/stamp.jsp?arnumber=1513512

29 P. A. Sikora, W. Platt, and J. Iannotti, "A new generation of EMI receiver system," in *IEEE International Symposium on Electromagnetic Compatibility, 1994, Symposium Record. Compatibility in the Loop*, pp. 132–137. Available at http://ieeexplore.ieee.org/stamp/stamp.jsp?arnumber=385669

30 A. Kempski and R. Smolenski, "Method of selection of dv/dt for EMI current ringing attenuation," in *Compatibility in Power Electronics, 2007 (CPE'07)*, 2007, pp. 1–5. Available at http://ieeexplore.ieee.org/stamp/stamp.jsp?arnumber=4296537

31 W. Teulings and B. Vrignon, "Fast conducted EMI prediction models for smart high-side switches," in *2011 8th Workshop on Electromagnetic Compatibility of Integrated Circuits (EMC Compo)*, Nov. 2011, pp. 159–164.

32 J. Luszcz, "Broadband modeling of motor cable impact on common mode currents in VFD," in *2011 IEEE International Symposium on Industrial Electronics (ISIE)*, IEEE, 2011, pp. 538–543. Available at http://ieeexplore.ieee.org/xpls/abs_all.jsp?arnumber=5984215

3

Conducted Emission Generation by Frequency Converter in ASD

It requires a much higher degree of imagination to understand the electromagnetic field than to understand invisible angels.

Richard Feynman

3.1 Classic Modulation Patterns Used in ASD

Pulse-width modulation (PWM) methods have been widely used in ASD applications for many years. There are also known and still developing other modulation methods used for optimizing particular performance parameters of ASD, for example, minimizing switching losses, minimizing output current harmonic distortions, extending linear modulation range, and simplifying implementation of modulation algorithms and others [1–3].

Conducted emission generation is also one of the important issues in ASD that can be optimized by modifying parameters of modulation strategy, by, for example, minimizing a number of transistor switching or minimizing common mode component of output voltages at the motor terminals [4–8]. In early developments of carrier-based PWM methods, sinusoidal waveforms of modulating signals have been used. Nowadays many other more advanced methods, for example, space-vector method, are used increasingly. To simplify further analysis of conducted emission generation and propagation in ASD, only classic, carrier-based PWM methods will be considered.

A typical graphical representation of space-vector modulation method used in three-phase two-level voltage source inverter is presented in Figure 3.1. According to this representation, it is possible to obtain eight unique sets of transistor switches states of converter's output bridge that allow obtaining six active voltage vectors $\vec{V}_1 - \vec{V}_6$ and two zero vectors $\vec{V}_0$ and $\vec{V}_7$.

High Frequency Conducted Emission in AC Motor Drives Fed by Frequency Converters: Sources and Propagation Paths, First Edition. Jaroslaw Luszcz.

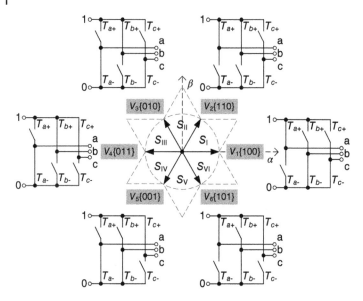

Figure 3.1 Inverter's transistor switching states in correlation with output voltage vector.

$$\vec{V}(t) = \sum_{i=0}^{i=7} \frac{t_i}{T_{PWM}} \vec{V}_i \tag{3.1}$$

Theoretically, all requested output voltage vectors at given time t can be obtained as composition of all accessible vectors using the formula (3.1) where $\vec{V}_i$ are component vectors and t_i are turn-on times of each of them. The sum of turn-on times of all vectors $\sum_{i=0}^{i=7} t_i$ is equal to modulation period T_{PWM}. There are many different equivalent vector compositions possible that allow achieving each of the required output voltage vectors. In practice, in frequency inverters, in order to reduce the excessive number of transistor switching processes, more simple and efficient method is used that allows composing required output voltage vectors using only two active vectors adjacent to the sector in which the resultant vector is expected. These sectors are indicated at space vector plane as S_I to S_{VI} in Figure 3.1. According to this assumption, each required output vector located in a given sector can be composed from only two active vectors adjacent to the particular sector. For example, output vectors with phase shift $0 < \varphi < \pi/3$ located in sector S_I can be composed of active vectors $\vec{V}_1$ and $\vec{V}_2$ and zero vectors $\vec{V}_0$ and $\vec{V}_7$ (3.2).

$$\vec{V} = \frac{t_1}{T_{PWM}} \vec{V}_1 + \frac{t_2}{T_{PWM}} \vec{V}_2 + \frac{t_0}{T_{PWM}} \vec{V}_0 + \frac{t_7}{T_{PWM}} \vec{V}_7 \tag{3.2}$$

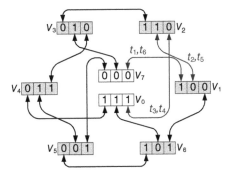

Figure 3.2 All possible state changes in three-phase bridge controlled according to the strategy of only one converter's leg transition.

The use of vectors adjacent to the given sector means that changing state of the transistor bridge is realized by changing state of one of the three inverter's legs only. It is evident that using switching vectors other than neighboring vectors will not be optimal from the point of view of, for instance, switching losses or conducted emission [9]. According to this minimum number of transistor switching strategy, all possible changes of three-phase bridge state are presented in Figure 3.2. Each change of bridge state shown in this diagram needs transition of only one leg of the bridge. Therefore, according to this rule, for each active state of the bridge $\vec{V}_i$, only two adjacent active vectors $\vec{V}_{i-1}$, $\vec{V}_{i+1}$ and one zero vector $\vec{V}_0$ or $\vec{V}_7$ are accessible.

Exemplary detailed graphical representation of one modulation period for output vector located in sector S_I for sinusoidal modulation signals is presented in Figure 3.3. According to the "only one leg transition" rule, three modulation signals V_a, V_b, and V_c are compared to triangular carrier signal, and in this way the required switching moments for each of the switches are determined in order to obtain the expected output voltage vector. Thus, during each modulation period T_{PWM}, there is only six switching processes necessary to use, two for each leg of the three-phase bridge. Particular topologies of bridge configurations during these six switching sequences are presented in Figure 3.4.

Frequency inverter used in ASD is loaded by AC motor windings connected via the motor feeding cable. Assuming that the load of inverter is usually substantially symmetric with respect to each phase and ground, switching sequences in each output phase can be considered as being analogical. Therefore, conducted emission generation phenomena in each phase, which are mainly associated with circuitry interactions with ground, can also be considered as being analogical in each phase. Assuming the converter's load symmetry, the specified six types of switching sequences, from the point of view of ground interactions, in terms of conducted emission generation, can be classified into only two categories:

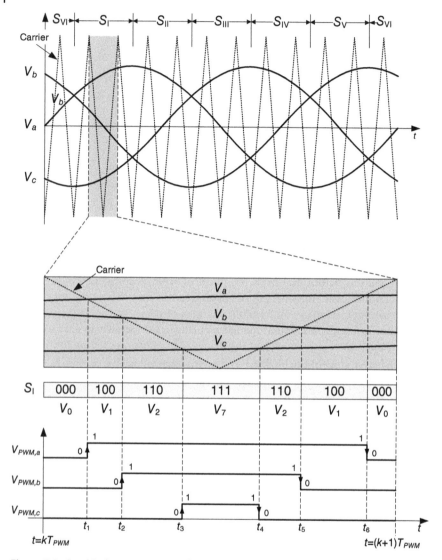

Figure 3.3 Graphical representation of carrier-based PWM sinusoidal modulation pattern for output vector located in sector S_I.

- Active-to-active vectors switching (A–A), where one of the output phase is switched from DC+ to DC– or vice versa, while the two other phases are continuously connected to DC bus terminals: one to DC+ and the other one to DC– (Figure 3.4, case t_2 and t_5).

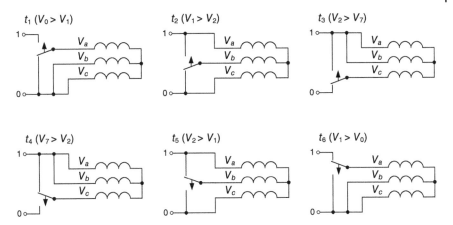

Figure 3.4 Three-phase bridge transitions used during one modulation period.

- Active-to-passive or passive-to-active vectors switching (A–P), where one phase is switched from DC+ to DC– or vice versa, while the two other phases are connected to the same DC bus terminal, DC+ or DC– (Figure 3.4, cases t_1, t_3 and t_4, t_6).

Summarizing, from the conducted emission generation point of view, the essential conclusion is that assuming a symmetry of converter's load, irrespective of the modulation method used, there are only two classes of transistor bridge state changes that result in characteristic disturbance effects. Therefore, generally conducted emission of ASD output inverter can be considered as superposition of conducted emissions correlated with these two transient classes, which are distributed in time domain according to the modulation pattern. This assumption allows significant simplification of models used for further analysis of conducted emission propagation in ASD with symmetrical load, which is very common in practice.

3.2 Conducted Emission Generation Pathways in ASD Applications

Fast commutations of switches in a frequency inverter necessary for achieving fundamental functionality of ASD, which is an appropriate control of low-frequency (LF) operating currents of AC motor, also cause rapid changes in voltages and resulting high-frequency (HF) currents [10,11]. Theoretically, HF voltages and currents generate high levels of electromagnetic emission associated with accompanied rapid changes of magnetic and electric field components, which can influence neighboring electrical devices by means of electric

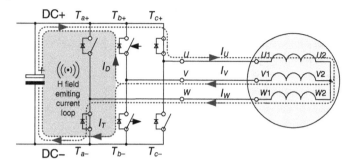

Figure 3.5 Example of inverter output current loop reconfiguration during commutation process.

and magnetic components of near field coupling. However, conducted emission of ASD depends primarily on effects associated with voltage steepness dv/dt during commutation [12,13]. In the commonly used topology of output side of frequency converter output bridge applied in ASD (Figure 3.5), commutated with high-speed currents, changes take place entirely only inside the frequency inverter bridge. This is a result of the use of freewheeling diodes integrated internally with switching transistors to reduce the loop's areas of commutated circuits and consequently inductances. Primarily, reduction of inductances of commutated circuits inside the output bridge is essential for limitation of over-voltages appearing during commutations and secondarily have very positive effect on limitation of generation of conducted emission originating from rapid current changes.

An example of modification of commutated current loop during typical commutation process is presented in Figure 3.5, where one of the converter's output terminal V is reconnected from DC− to DC+ bus voltage. During this commutation, the external inverter current I_V remains roughly unchanged because of relatively high inductance of the motor winding. The only change of commutated current loop takes place inside the inverter; current in transistor $T_{b−}$ is switched off and the same current is redirected to freewheeling diode of transistor T_{b+}.

During the same exemplary commutation process, the output voltage of converter terminal V is rapidly changed from DC− bus potential into DC+ bus potential. The essential difference, from the point of view of electromagnetic emission generation, is that this rapid change of voltage is transmitted outside the inverter for a relatively long distance, via one wire of the motor feeding cable toward the motor terminal and then to entire motor windings (Figure 3.6). This rapid voltage change along the motor cable and winding is a source of electric field transient that can be coupled capacitively with all neighboring conducting objects, like two other wires of the motor feeding cable, two other

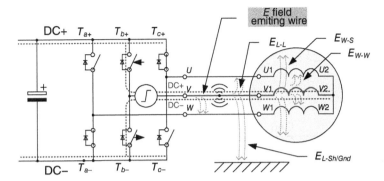

Figure 3.6 Example of converter's output voltage repolarization during commutation process.

motor windings, grounded motor stator, grounded shield of the motor cable and other, especially grounded conducting structures (Figure 3.6).

Detailed analysis of voltages and currents distribution at motor side of frequency inverter requires taking into account low-frequency phenomena related to motor speed control and simultaneously high-frequency phenomena that are side effects of the use of DC voltage modulation with high-speed switching devices. Therefore, particular definitions of voltage and current components appearing at inverter's load circuitry should be determined to clarify specific character of some of the analyzed effects.

Complete distribution of voltages at AC motor terminals with special focus on the ground reference can be fully described using three differential mode (DM) voltage components $V_{DM,ab}$, $V_{DM,bc}$, and $V_{DM,ac}$ determined for each pair of phases and three common mode (CM) voltage components $V_{CM,a}$, $V_{CM,b}$, and $V_{CM,c}$ determined for each phase with relation to ground (Figure 3.7). For analysis in the low-frequency range, close to power frequency, these DM and CM voltages are commonly known as line voltages V_{L1-L2}, V_{L2-L3}, V_{L3-L1} and phase voltages V_{L1}, V_{L2}, V_{L3}, respectively.

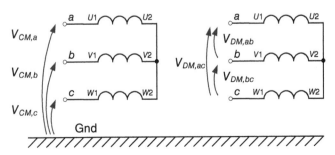

Figure 3.7 Common and differential mode voltage components in AC motor.

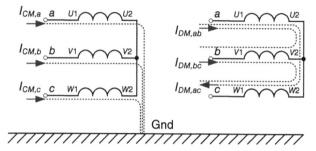

Figure 3.8 Common and differential mode current components in AC motor.

Similarly, differential mode currents I_{L1-L2}, I_{L2-L3}, I_{L3-L1} and common mode currents I_{L1}, I_{L2}, I_{L3} can be defined as is shown in Figure 3.8. Based on this definition, it is clear that in ASD characteristic CM currents at motor side flow through parasitic capacitive couplings between output side circuitry and ground.

3.2.1 DM Current Paths at Motor Side of Frequency Converter

Differential mode (DM) output currents of an FC can be divided into two spectral components: low-frequency DM currents and high-frequency DM currents. Low-frequency DM current components that flow trough the entire length of the motor windings have relatively high rms magnitudes, which generate primary component of magnetic field and subsequently motor torque at different rotor speeds controlled by output current frequencies. This phenomenon is the fundamental functionality of power conversion in AC motor drives.

High-frequency components of DM currents flowing in AC motor fed by frequency converter are induced by DM-voltage high-speed changes produced by frequency inverter during transistor switching transients [14]. High-frequency DM currents start to flow the same way as low-frequency currents, but along motor feeding cable, and motor windings are shunted by interwire parasitic capacitances of motor feeding cable $C_{(Wr-Wr)1} - C_{(Wr-Wr)n}$ and interwinding parasitic capacitances of motor windings $C_{(Wn-Wn)1} - C_{(Wn-Wn)n}$ (Figure 3.9) [15,16]. Parasitic capacitances of motor windings and wires of motor feeding cable are relatively small, for the most common cases in the range of nanofarad but large enough to form relatively low impedance for HF signals. Therefore, shunting impedances resulting from these capacitances determine HF currents distribution.

In most cases, values of high-frequency DM currents decrease with the distance from converter's terminals. Nevertheless, distribution inequality of parasitic capacitances along motor windings and feeding cable may lead to significant irregularity of high-frequency DM currents. Unequal distribution of

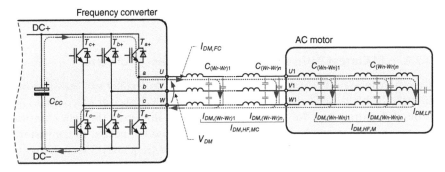

Figure 3.9 Simplified representation of DM currents distribution at motor side of frequency converter.

high-frequency DM currents can also be an effect of possible signal reflections occurring because of these inequalities. Simplified, graphical distribution of high-frequency CM current paths in a typical ASD is presented in Figure 3.9.

According to the presented graphical representation, parasitic capacitances of feeding cable wires and motor windings significantly affect high-frequency DM current components distribution along motor cable and windings. The total high-frequency DM current at FC output side is a sum of currents resulting from motor cable interwire parasitic capacitances $I_{DM,HF,MC}$ (3.3) and motor interwinding parasitic capacitances $I_{DM,HF,M}$ (3.4), whereas total DM current also contains a low-frequency component $I_{DM,LF}$ (3.5):

$$I_{DM,HF,MC} = \sum_{i-1}^{n} I_{DM,(Wr-Wr)i} \qquad (3.3)$$

$$I_{DM,HF,M} = \sum_{i-1}^{n} I_{DM,(Wn-Wn)i} \qquad (3.4)$$

$$I_{DM,FC} = I_{DM,HF,MC} + I_{DM,HF,M} + I_{DM,LF} \qquad (3.5)$$

For a given single frequency, distribution of DM currents along motor windings is decreasing lengthways motor cable and motor windings; therefore, higher amplitudes appear generally closer to FC terminals and closer to motor winding terminals from the point of view of AC motor. With the decrease of frequency of spectral components of DM currents, their pathway changes, they penetrate deeper and deeper into windings, and finally cover the whole winding at low frequencies for which parasitic capacitances can be neglected. In ASD applications where transmission line effects and reflections at motor terminals and inside motor windings become significant, high-frequency DM current distribution is more complex and difficult to analyze using simplified models.

3.2.2 CM Current Paths at Motor Side of Frequency Converter

Common mode HF currents in AC motor fed by frequency converter are induced by high-speed changes of voltages between converter's terminals and ground. Contrary to DM currents, CM currents flow only through parasitic capacitances to ground of frequency converter and connected to converter's load—AC motor with feeding cable. Therefore, low-frequency CM currents are very low, and can be neglected in analysis.

CM parasitic capacitances between converter's load components and ground are usually significantly higher than corresponding interwire and interwinding DM capacitances of the same ASD. With the increase of frequency, impedances of parasitic couplings between motor windings and grounded motor frame become smaller; therefore, high-frequency components of CM currents achieve significantly higher magnitudes.

Furthermore, CM currents in relation to DM currents can be potentially more disturbing because they flow through neighboring grounded components and its flow paths are more difficult to predict and control [17,18]. The above characteristic features of CM currents result in that they are usually the main issue related to conducted emission of ASD application [11]. An example of CM currents distribution in typical ASD application is presented in Figure 3.10.

Based on this graphical representation of CM current distribution, it can be noticed that total CM current generated at output side of frequency converter can be considered as a sum of CM current of motor winding $I_{CM,MF}$ and CM current $I_{CM,Sh}$ of the motor feeding cable (3.6). Both of these CM currents

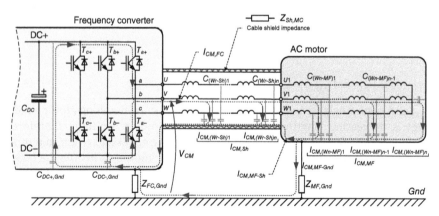

Figure 3.10 Simplified representation of CM currents distribution at motor side of frequency converter.

$I_{CM,MF}$ and $I_{CM,Sh}$ can be described as a sum of current subcomponents correlated with distributed CM parasitic capacitances of motor windings $C_{(Wn-MF)1}$
$- C_{(Wn-MF)n}$ (3.7) and the motor feeding cable $C_{(Wn-Sh)1} - C_{(Wn-Sh)n}$ (3.8).

$$I_{CM,FC} = I_{CM,Sh} + I_{CM,MF} \qquad (3.6)$$

$$I_{CM,MF} = \sum_{i-1}^{n} I_{CM,(Wn-MF)i} \qquad (3.7)$$

$$I_{CM,Sh} = \sum_{i-1}^{n} I_{CM,(Wr-Sh)i} \qquad (3.8)$$

As for DM currents, CM current distribution depends on distribution of
parasitic CM capacitances between cable wires and ground, and windings
and ground. Furthermore, CM currents depend additionally on grounding
impedances of motor $Z_{MF,Gnd}$ and frequency converter $Z_{FC,Gnd}$, which are
the key parameters for controlling conducted and radiated emission of ASD.

According to the above simplified representation of CM current distribution
(Figure 3.10), different return paths of CM current of motor frame $I_{CM,MF}$
are possible: through cable shield impedance $Z_{Sh,MC}$ and through motor
grounding impedance $Z_{MF,Gnd}$ and frequency converter grounding impedance
$Z_{FC,Gnd}$. This means that impedance of the motor cable shied has extra specific influence on CM currents distribution in ASD and therefore on resultant
conducted emission. More detailed analysis of significance of motor cable parameters on conducted emission of ASD is presented in the following sections.

3.3 Mutual Coexistence of CM and DM Voltages and Currents in ASD

Classification of high-frequency parasitic currents in ASD into DM and CM
has been developed by many researchers and presented in Sections 3.2.1 and
3.2.2 [19–21]. This approach allows simplifying analysis significantly, especially
when only one of the current components is highlighted: CM or DM [22,23].

The main issue with the analysis of HF parasitic current paths in ASD is
related to the fact that DM and CM currents usually exist simultaneously and
flow partly via the same paths, and therefore are substantially mixed. In real
ASD systems, there is usually a number of load circuit asymmetries, especially
in HF range, due to which DM and CM components are transformed into each

other [24,25]. Thus, it is difficult to measure separately these two types of CM currents in experimental evaluations.

Some measurement methods that allow measuring CM and DM parasitic current components separately exist [26–28], although these methods have a number of limitations and are focused primarily on measurement of external conducted emission of devices into power grid, according to CISPR standard recommendations in frequency domain by using EMI receiver. Analysis of internal distribution of DM and CM currents inside frequency converters require different, more complex approach. This is one of the key difficulties faced in RFI filter design process [22,29].

Standardized EMI measurement in frequency domain is usually focused only on amplitudes of spectrum components; therefore, the lack of phase information significantly limits the analysis of detailed CM and DM currents distribution in relation to time domain analysis capabilities. On the other hand, time domain measurements of HF components of currents and voltages in ASD, which are usually much lower than LF components, require using measurement equipment with very high measuring dynamic range. This measurement difficulty usually leads to the obligation of using high-pass filters to separate HF components from LF components of voltages and currents, which significantly impede measurement accuracy. However, time domain measurement approach is more convenient for detailed identification of specific current paths for a selected components of ASD [30–32].

3.3.1 CM-to-DM Conversion of Voltages and Currents in ASD

Power transfer in three-phase power systems is commonly implemented as balanced with local reference ground, which is usually connected to earth via protective earth terminal (PE) of the power grid. The fundamental reason for generating CM voltages and currents in the initially balanced three-phase power system is an unbalance of load impedances. Any CM current flowing in power system is highly undesirable and usually leads to some adverse effects. Unequal distribution of linear loads between phases is known as a source of zero-sequence current component, which is very inadvisable in power systems [22,33].

This kind of unbalance, which can be named as static unbalance, generate low-frequency components of CM voltages and currents at frequencies correlated to fundamental frequency of power grid. Much more convoluted effects in power system can cause dynamic kind of load unbalance, which takes place intensively in nonlinear load, where impedance is changing significantly and repeatedly with frequency much higher than power grid frequency, which is usually the modulation frequency. This kind of unbalance is a fundamental source of HF components of CM voltages and currents injected into the power grid, which is the foremost origin of conducted and radiated electromagnetic emission. Only fully symmetric power transfer via perfectly balanced transmission line allows avoiding CM voltages generation entirely.

Generally, full symmetry of impedances is difficult to achieve in three-phase power circuits and almost impossible in power electronic converters that are nonlinear and their impedances change exceptionally and rapidly during transistors switching [24]. Therefore, in power electronic devices, CM voltages and currents are intensively generated, and because of the existing unbalances of internal and external impedances, it can be converted into DM and vice versa [24,33].

In practical circuitry, most of components, especially three-phase components, expose some unbalance of impedances in relation to ground, more specifically to neighboring components that are grounded. These asymmetries are an effect of the commonly used configuration of three-phase power systems where many components, for example, three-phase transformers, and power lines are constructed as slightly imbalanced between phases and therefore also between phases and ground. Another evident reason of power systems unbalance is the use of one-phase powered devices.

In the low-frequencies range, close to fundamental frequency of power system, asymmetries are relatively small, usually below 1%, depending on the rated power of analyzed system. Nevertheless, with the increase of frequency, asymmetries of impedances usually become more significant, because of asymmetrical distribution of unavoidable parasitic coupling between components. Commonly appearing parasitic capacitive and inductive couplings usually behave linearly in most cases. Therefore, this kind of imbalance of HF impedance of power electronic circuitry can be named as linear unbalance that can significantly influence conducted emission propagation—especially in frequency range above hundreds of kilohertz.

In a simplified representation of ASD application adequate for limited frequency range, most of CM and DM impedances can be identified and described as individual equivalent components by using lumped parameters (Figure 3.11): $Z_{DM,FC}$—differential mode phase impedance of frequency converter, $Z_{CM,ASD}$—common mode phase impedance of whole ASD (frequency converter, AC motor, and motor cable), and $Z_{DM,M}$—differential mode phase impedance of AC motor.

In Figure 3.11 a highly simplified representation of CM currents distribution in ASD powered from three-phase power grid is provided. To simplify the explanation of CM-to-DM voltage cross-conversion in ASD, in the presented circuit CM voltage source $V_{CM,PG}$ has been introduced into the analyzed system at grid side. This CM voltage source results in three CM currents $\vec{I}_{CM,L1}$, $\vec{I}_{CM,L2}$, and $\vec{I}_{CM,L3}$, flowing in all phases via DM equivalent inductances of power grid $Z_{DM,PG}$, frequency converter $Z_{DM,FC}$, and an equivalent CM impedance of the whole ASD $Z_{CM,ASD}$.

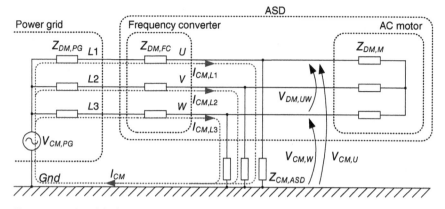

Figure 3.11 Simplified representation of CM-to-DM voltages conversion in ASD.

Assuming that adequate impedances are unbalanced, thus not equal for each phase, CM currents in each phase will also be not equal and thus generate corresponding differential voltages between all phases at motor side, for example, $V_{DM,UW}$ for phases U and W (3.9):

$$\vec{V}_{DM,U-W} = \vec{V}_{CM,U} - \vec{V}_{CM,W} \tag{3.9}$$

The conversion factor of grid-side CM voltage $V_{CM,PG}$ into motor-side CM voltage $V_{CM,U}$ determined for each phase can be defined as (3.10) where DM impedance Z_{DM} for each phase is defined as a sum of DM impedance of power grid $Z_{DM,PG}$ and DM impedance of frequency converter $Z_{DM,FC}$ (3.11).

$$\vec{V}_{CM,U} = \vec{V}_{CM,PG} \frac{\vec{Z}_{CM,U}}{\vec{Z}_{DM,U} + \vec{Z}_{CM,U}} \tag{3.10}$$

$$\vec{Z}_{DM,U} = \vec{Z}_{DM,PG,U} + \vec{Z}_{DM,FC,U} \tag{3.11}$$

Based on these equations the CM-to-DM voltage transfer factor $T_{CMV \rightarrow DMV}$ can be defined as (3.12). According to this formula, differences in ratio between CM and DM impedances Z_{DM}/Z_{CM} for each pair of phases are the main factor influencing DM components generation at motor side of ASD due to the existence of CM component in the power grid.

$$T_{CMV \rightarrow DMV} = \frac{V_{DM,U-W}}{V_{CM}} = \frac{1}{\frac{Z_{DM,U}}{Z_{CM,U}} + 1} - \frac{1}{\frac{Z_{DM,W}}{Z_{CM,W}} + 1} \tag{3.12}$$

The CM impedance of output side of FC is usually capacitive and results mainly from parasitic capacitances of motor windings and motor cable, because in ASD generally there is no galvanic connections between motor windings and ground. Magnitudes of impedances of capacitive couplings depend on frequency, therefore assuming fixed unbalance level of parasitic capacitances between phases; CM-to-DM transfer factor changes considerably with frequency and becomes more meaningful in the upper frequency range.

3.3.2 DM-to-CM Conversion of Voltages and Currents in ASD

DM voltages and currents in ASD, in contrast to CM, are much higher, depending on the transferred power–thus higher for higher power transferred. In the power grid, small voltage asymmetry usually occurs, resulting mainly from load unbalance between phases and internal asymmetries of impedances of many of powered devices. The acceptable voltage unbalance level defined by power quality standards is 2–3% depending on the type of environment.

In power systems, there are many known adverse reactions and harmful effects of neutral wire currents, which can be classified as specific, low-frequency type of CM currents, otherwise known as zero-sequence currents. Commonly observed reasons for generating low-frequency CM currents in power grids are related to existing unbalance of the power grid and harmonic distortions of load currents. Especially harmonic components on orders of multiple of three can directly lead to undesirable neutral current flow [34,35]. The essential difference between zero-sequence currents and CM currents correlated with conducted emission generation is that they are of much lower frequencies and flow primarily through real galvanic earthing connections, like neutral wires that are intentionally installed for carrying this kind of currents. Simplified circuit model of three-phase power system that represents DM-to-CM voltage conversion is presented in Figure 3.12.

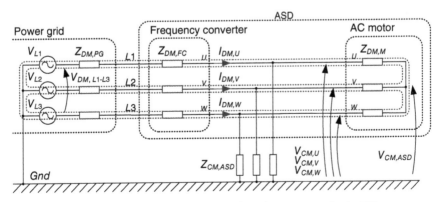

Figure 3.12 Simplified representation of DM-to-CM voltages conversion in ASD.

Assuming that DM impedances of power grid $\vec{Z}_{DM,PG}$, frequency converter $\vec{Z}_{DM,FC}$, and motor $\vec{Z}_{DM,M}$ are usually much smaller than total equivalent CM impedance of whole ASD $\vec{Z}_{CM,ASD}$ within LF range, the CM voltage generated due to grid voltage unbalance and DM impedances asymmetries can be expressed as (3.13) according to Figure 3.12:

$$\vec{V}_{CM} = \vec{V}_{NN'} + \vec{V}_{N'N''} \tag{3.13}$$

where $\vec{V}_{NN'}$ and $\vec{V}_{N'N''}$ are CM voltages related to asymmetry of voltage of power grid (3.14) and asymmetry of DM impedances (3.15), respectively.

$$\vec{V}_{NN'} = \vec{V}_{L12} + \vec{V}_{L23} + \vec{V}_{L31} \tag{3.14}$$

$$\vec{V}_{N'N''} = \vec{V}_{CM,U} + \vec{V}_{CM,V} + \vec{V}_{CM,W} \tag{3.15}$$

CM voltages at AC motor terminals $\vec{V}_{CM,U}$, $\vec{V}_{CM,V}$, $\vec{V}_{CM,W}$ can be determined as a difference of grid voltages and adequate voltage drops due to phase currents $\vec{I}_{DM,U}$, $\vec{I}_{DM,V}$, $\vec{I}_{DM,W}$ and adequate impedances of power grid $\vec{Z}_{DM,PG}$, frequency converter $\vec{Z}_{DM,FC}$, and AC motor $\vec{Z}_{DM,M}$ (3.16).

$$\vec{V}_{CM,U} = \vec{V}_{L1} - \vec{I}_{DM,U} \left(\vec{Z}_{DM,PG,L1} + \vec{Z}_{DM,FC,L1} + \vec{Z}_{DM,M,L1} \right) \tag{3.16}$$

Hence, it follows that in steady-state condition, when transistor switching is not considered, CM voltage $\vec{V}_{CM,ASD}$ is a result of unbalance of entire phase impedances $\vec{Z}_{DM,U}$, $\vec{Z}_{DM,V}$, and $\vec{Z}_{DM,W}$ and unbalance of phase currents $\vec{I}_{DM,U}$, $\vec{I}_{DM,V}$, and $\vec{I}_{DM,U}$. Total DM impedance of each phase $\vec{Z}_{DM,U}$, $\vec{Z}_{DM,V}$, and $\vec{Z}_{DM,W}$ is a sum of adequate impedances of power grid $\vec{Z}_{DM,PG,U}$, frequency converter $\vec{Z}_{DM,FC,U}$, and AC motor $\vec{Z}_{DM,M,U}$ (3.17).

$$\vec{Z}_{DM,U} = \vec{Z}_{DM,PG,U} + \vec{Z}_{DM,FC,U} + \vec{Z}_{DM,M,U} \tag{3.17}$$

The DM-to-CM voltage transfer factor $T_{DMV \to CMV}$ can be defined as (3.18). According to this formula, differences between DM impedances between phases are the main factor influencing DM-to-CM transfer factor.

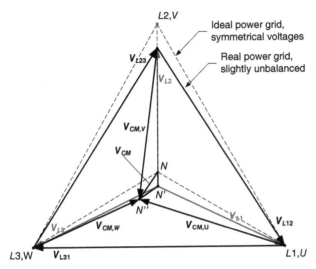

Figure 3.13 CM voltage components resulting from voltage and impedance asymmetries occurring in three-phase power system.

$$T_{DMV \rightarrow CMV} = \frac{V_{CM}}{V_{DM,U-W}} = \frac{1}{1 + \dfrac{Z_{DM,W}}{Z_{DM,U}}} \tag{3.18}$$

An example of graphical representation of movement of neutral point potential from $\vec{N}$ to $\vec{N'}$, due to the powder grid unbalance, and further from $\vec{N'}$ to $\vec{N''}$, due to DM impedance asymmetries of analyzed load, together with power grid impedance is presented as the vector diagram in Figure 3.13.

Summarizing, CM voltages occurring in ASD depend on unbalance of grid voltage and phase impedances. In a power grid environment with dominating linear loads, voltage unbalance is usually maintained at the low level, below few percent; therefore, the levels of CM voltage, possible to appear, are usually not significant. Problem of generation of CM voltage components increases significantly when nonlinear loads are used, for example, ASD, due to meaningful and frequently appearing changes of impedances of frequency converter's circuitry.

Rough analysis of commutation processes in ASD, diodes commutation at grid-side rectifier, and transistors commutation at motor-side inverter show that from the point of view of the power grid, most of the time ASD as a load is seen as only two-phase load, thus enormously asymmetric. Similarly, instantaneous symmetry at motor-side frequency converters cannot be achieved, because AC motor windings as a three-phase circuitry are fed by switched DC bus voltage, thus highly asymmetrically (Figure 3.14).

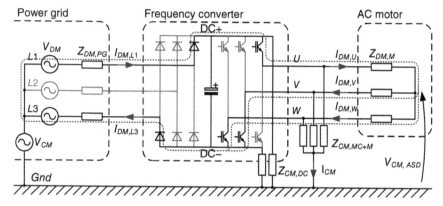

Figure 3.14 Instantaneous asymmetry of line-side and motor-side currents of frequency converter in ASD.

Instantaneous unbalance caused by commutation processes inside FC can be named as dynamic unbalance and its significance for CM voltage generation in ASD is essential. Nevertheless, unbalance resulting from static asymmetry of ASD impedances described above also increases additionally final CM voltages generated by dynamically switched impedances.

According to Figure 3.14, assuming that $I_{DM,L1} = I_{DM,L3}$ and $I_{DM,L2} = 0$, grid-side current unbalance, calculated as a percentage proportion between maximum deviation from average current to average current according to simplified definition ANSI/IEEE (3.19) achieves the level of 100%. Line voltage unbalance at motor side of FC achieves the level of 50%.

$$\text{Unbalance}(\%) = \frac{\text{Maximum Deviation from Average}}{\text{Average of Three Phases}} \, 100\% \qquad (3.19)$$

To conclude, dominating source of unbalance in ASD are commutations of internal subcircuits due to typical functionality of switching devices like transistors and diodes. During typical commutation process, significant parts of a circuit are usually rapidly reconnected, thus associated impedances also change, causing a change of total symmetry. These dynamic changes of symmetry are the main origin of CM voltages that are the main cause of conducted emission generation in ASD.

3.3.3 Motor-Side DM Currents Distribution During Active-Vectors Switching

Switching between two active vectors (A–A) in voltage source inverter (VSI), cases t_2 and t_5 presented in Figure 3.4, is a sequence where one of the bridge

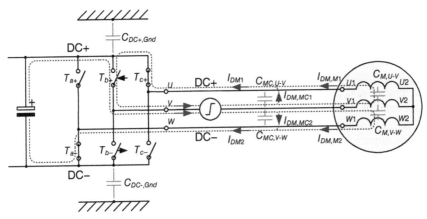

Figure 3.15 DM currents distribution at converter's output side during active vectors switching.

legs changes potential from one of the DC bus line to the opposite one. It can be realized by switching two transistors, for example, by opening T_{b-} and closing T_{b+} according to Figure 3.15. The presented active vectors switching sequence results in voltage change at the inverter terminal V from V_{DC-} to V_{DC+}, which initiate two transient processes:

- One that is expected to change the output current in phase I_V, which means an increase of current value if it was positive or a decrease if it was negative before switching,
- The other that is tightly correlated to reloading of interwire parasitic capacitances of motor cable $C_{MC,V-W}$, $C_{MC,V-U}$ and interwinding parasitic capacitances of motor windings $C_{M,V-W}$, $C_{M,V-U}$, which generate stray high-frequency DM current transients.

The first transient process is intended and allows controlling the motor current according to PWM control requirements. Motor currents can change relatively slowly because motor inductance is relatively high. The second transient process is not intended and initiated by rather small parasitic capacitances that are distributed along motor cable wires and motor windings, therefore connected almost directly, via low inductance of wire connections, to converter's output. This process is high-speed transient, because after each output voltage change parasitic capacitances are reloaded to the opposite polarity via much smaller than motor's inductance and thereby high values of very short current pulses are generated. This type of high-frequency currents are the main origin of conducted emission.

From the viewpoint of conducted emission generation, the flow paths of high-frequency current transients are essential. The two main loops of HF currents accompanied with high-speed reloading processes are presented in Figure 3.15. One of them, which can be named as charging loop, is represented by DM current $I_{DM,MC2}$ and charge differential parasitic capacitances between phases V and W of motor cable wires $C_{MC,V-W}$ and motor windings $C_{M,V-W}$ up to the voltage of DC bus. Second loop of HF parasitic current is represented by discharging current $I_{DM,MC1}$. In this loop differential parasitic capacitances between phases U and V of motor cable wires $C_{MC,U-V}$ and motor windings $C_{M,U-V}$, earlier charged up to the voltage of DC bus, are discharged via low impedance of switched-on T_{b+} transistor.

The complete path of differential current loops $I_{DM,MC1}$ and $I_{DM,MC2}$ are presented in Figure 3.15. It can be noticed that charging energy is taken from DC bus capacitor and discharging energy is dissipated in switching-on transistor T_{b+} and resistances of motor windings and feeding cable. Based on currents distribution in Figure 3.15, it can also be noticed that discharging current $I_{DM,MC1}$ and charging current $I_{DM,MC2}$ are added together in switched phase V and in transistor T_{b+}. Summarizing, during output voltage switching, the highest high-frequency DM current is injected into the commutated phase and returned back to the converter via two other phases not always evenly distributed.

3.3.4 Motor-Side CM Currents Distribution During Active Vectors Switching

CM current components are generated at output side of frequency converter simultaneously with DM currents and flow through parasitic capacitances between grounded parts of ASD and energized motor cable wires and motor windings. In typical applications, the most meaningful grounded components from the point of view of converter's output-side CM currents are AC motor stator and motor cable shield. Before analogous switching process as described in previous Section 3.3.3, where T_{b-} is opened and T_{b+} is closed, parasitic capacitances of switched phase of motor cable wire related to ground $C_{MC,V-Gnd}$ and motor winding $C_{M,V-Gnd}$ are precharged up to V_{DC-} voltage, which is negative in relation to ground. After vector switching, these two types of equivalent parasitic capacitances are reloaded to the opposite voltage V_{DC+}. The complete current path of CM current I_{CM}, which is accompanying this reloading process, is presented in Figure 3.16.

Generated CM current I_{CM} flows through the DC bus capacitor, motor cable capacitance $C_{MC,V-Gnd}$, motor winding capacitance $C_{M,V-Gnd}$, DC− bus to ground parasitic capacitance $C_{DC-,Gnd}$, and grounding connections between motor and frequency converter. Effects of these parasitic CM currents are substantially different and more essential for EMI generation than adequate DM currents. First, CM currents flow through grounding connections between motor and converter, and thus can spread out outside ASD by means of grounded

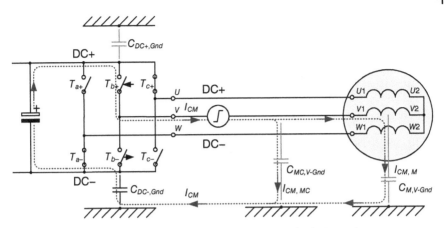

Figure 3.16 CM currents distribution at converter's output side during active vectors switching.

structural elements. Therefore, their paths cannot be explicitly defined and depend on impedances in high frequency range of external grounding connections. Second, CM currents significantly depend on DC bus parasitic capacitances $C_{DC+,Gnd}$ and $C_{DC-,Gnd}$ in relation to ground, which form the essential parasitic coupling path between motor side and grid side of frequency converter. Summarizing, CM currents generated at motor side of frequency converter are the main origin of conducted emission in ASD.

3.4 Experimental Study of Inverter's Output Current and Voltage Waveforms

Distribution of high-frequency components of voltages and currents during transistor switching can be confirmed based on experimentally recorded waveforms of phase voltages and accompanied HF current transients at FC output side. Accurate measurement of high-frequency transient currents at inverter's output require using high-pass filtered current probes to minimize influence of current components of low frequency related to motor speed and high magnitude related to motor load on measurement resolution. Another method to overcome difficulties with coexistence of LF and HF current components is the use of recording equipment with extra high-amplitude resolution and current probes protected against saturation of magnetic core by low-frequency high-magnitude motor currents.

These two techniques help limit disadvantageous influence of relatively high-magnitude low-frequency DM current components on the maximum measurement resolution possible to obtain in high frequency range. In Figure 3.17, an

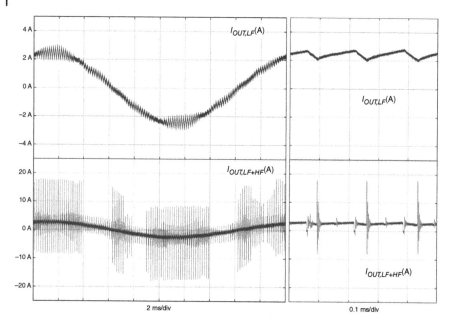

Figure 3.17 Low- and high-frequency components of output currents of frequency converter.

example of recorded output current of one phase of frequency converter is presented. In the bottom waveform, total output current $I_{OUT,LF+HF}$ of FC is presented, which include LF and HF components. The recorded peak current values rise up to almost 20 A, whereas the amplitude of the motor load current of output frequency of 50 Hz is below 3 A. In the upper waveform, motor-phase current $I_{OUT,LF}$ is presented, including LF component in which only current ripples of frequency of PWM modulation are visible.

3.4.1 Voltage Transients at AC Motor's Terminals

An example of characteristic phase-to-ground voltage waveforms observed at terminal of AC motor windings fed by FC is presented in Figure 3.18. The marked single period of PWM modulation T_{PWM} contains six transistor's switching sequences $t_1 - t_6$, where two of them t_2 and t_5 are switchings between active vectors (A–A) and the other cases t_1, t_3 and t_4, t_6 are switchings between active-to-passive or passive-to-active vectors (A–P).

In the presented example, visual comparison of voltage ringing waveforms during switching between active–active and active–passive vectors shows no significant differences in generated HF voltage ringing magnitudes and frequencies. It indicates that the expected conducted emission as an effect of all

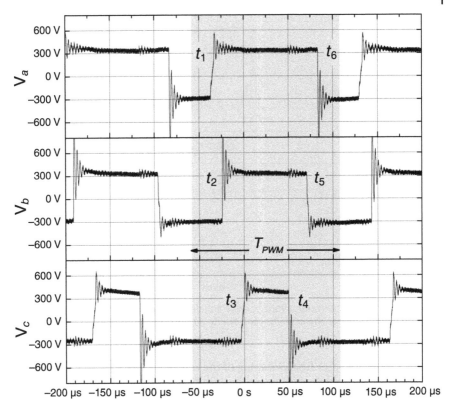

Figure 3.18 Example of experimentally obtained three-phase voltages at terminals of AC motor fed by frequency converter.

switching processes should be rather similar and not so much dependent on the type of switched vectors (A–A) or (A–P).

On the other hand, on all three voltage waveforms every other switching voltage slope is much longer than the other, for example, slope t_6 is about 0.5 μs long, whereas slope t_1 lasts much longer, about 4 μs. Adequate waveforms in the other two phases have very similar relationship. Assuming that impedance of converter's load is relatively symmetrical—as it was stated in Section 3.1—there is no topological reason for such a significant difference in the switching processes t_1 and t_6, t_2 and t_5, and t_3 and t_4. The explanation for this phenomenon can be found by correlating voltage switching waveforms with current waveforms in the switched phases.

In Figure 3.19, two similar voltage switching patterns recorded for time periods where the direction of output current was different are presented. In the upper set of voltage and current waveforms, voltage is switched from V_{DC+} to V_{DC-} while the value of commutated current is negative and its absolute value

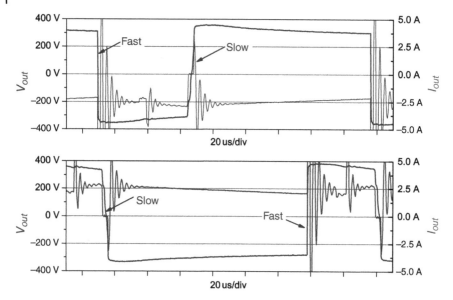

Figure 3.19 Converters output voltage switching times during modulation of negative output currents (upper characteristic) and during modulation of positive output currents (lower characteristic).

starts increasing after first switching of the transistor marked as "fast" and can be named as switching-current-on pattern. The output voltage switching slope in this type of commutation is faster than for the opposite voltage switching from V_{DC-} to V_{DC+}, marked as "slow" where the generated slope of output voltage takes much longer time. In the lower set of waveforms, where the value of commutated current is positive, the same voltage change from V_{DC+} to V_{DC-} results in the decrease of output current absolute value (switching-current-off pattern) and therefore the resultant voltage slope is significantly lower than for the opposite voltage switching.

Summarizing, in each phase of output voltage of inverter, every second voltage switching that initiate an increase of an absolute value of output current in adequate phase is slow. Switching times in these cases depend mainly on switching characteristic of the transistor, because the process of switching-on of the adequate transistor results in the immediate connection of commutated output phase with DC bus voltage.

The opposite switching patterns, switching-off of the adequate transistor, which start process of decreasing of an absolute value of commutated current, results in different commutation process that is relatively slower voltage switching. Switching time of these types of commutation processes is not directly related to transistor switching process, because commutated current is supported by relatively large inductance of motor winding and cannot decrease

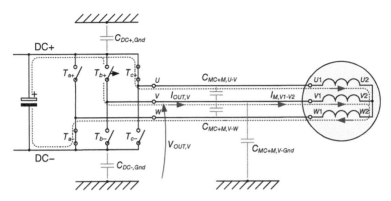

Figure 3.20 Converter's output current distribution at the beginning of slow commutation process and lumped representation of CM and DM parasitic capacitances of motor cable and motor windings.

immediately. This commutation type is presented in Figure 3.20, where converter's output currents distribution just before switching-off of the transistor T_{b+} is shown.

Analysis of recorded output currents during slow type of commutation process (Figure 3.21) clarifies that converter's output currents $I_{OUT,V}$ after transistor switching-off (transition between time intervals a and b) achieve zero value immediately, whereas motor current $I_{M,V1-V2}$ in the same phase still continue flowing at similar level.

During the time interval marked as b, commutated voltage is decreasing nearly linearly to achieve the opposite voltage of DC bus, that is, V_{DC-}, and then output current $I_{OUT,V}$ starts increasing again from zero value to achieve approximately previous value after short ringing process (time interval c). To explain temporary inconsistency in converter's output current $I_{OUT,V}$ and motor current $I_{M,V1-V2}$, let us assume that all distributed parasitic capacitances of motor cable and motor winding of the commutated phase are considered as lumped and both are connected in presented circuit (Figure 3.20) to motor cable. As a result of this assumption, the motor winding can be considered as inductance only without parasitic capacitances.

Equivalent lumped DM capacitances $C_{MC+M,U-V}$ and $C_{MC+M,V-W}$ presented in Figure 3.20 represent sums of adequate capacitances of motor cable $C_{MC,U-V}$, motor windings $C_{M,U-V}$ between phases U and V (Figure 3.15), and $C_{MC,V-W}$, $C_{M,V-W}$, respectively, between phases U and V. Adequately, lumped equivalent CM capacitance $C_{MC+M,V-Gnd}$ represents sum of capacitances of motor cable $C_{MC,V-Gnd}$ and motor windings $C_{M,V-Gnd}$ (Figure 3.16) between phase V and ground.

Parasitic capacitances of the commutated phase V, specified in Figure 3.20, are essential for detailed analysis of slow commutation process that takes place

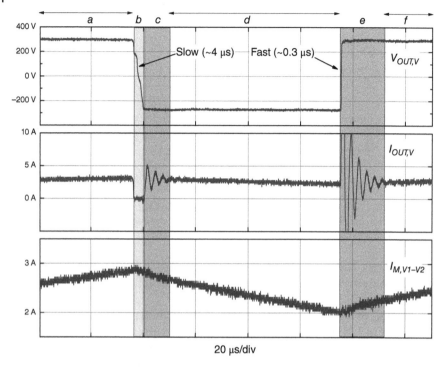

Figure 3.21 Converter's output voltage and current waveforms during slow and fast commutation processes.

after switching-off of the transistor T_{b+}. Output current $I_{OUT,V}$, which earlier was flowing through T_{b+} and motor winding $V1 - V2$, must continue flowing because energy stored in the motor inductance will support it. The current loop that finally allows keeping this current flowing can be formed with the use of freewheeling diode integrated with the transistor T_{b+}, motor windings $V1 - V2$ and $W2 - W1$, and the transistor T_{b+}. It is commonly known description of commutation process of output currents in frequency converter feeding AC motor when parasitic capacitances do not have to be considered.

However, before freewheeling diode is able to take over the commutated current $I_{OUT,V}$, the voltage at output terminal V must decrease to the level close to V_{DC-}. To achieve this state, some of the parasitic capacitances of motor cable and motor windings must be recharged, especially $C_{MC+M,V-W}$ must be discharged, because before switching-off was charged up to V_{DC+} and finally should be connected with V_{DC-}. During the analyzed time period b (Figure 3.21), capacitance $C_{MC+M,V-W}$ is connected in parallel with the freewheeling diode integrated with the transistor T_{b-} via switched-on transistor T_{a-}, thus reversely

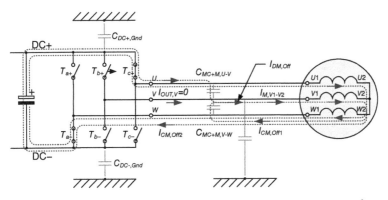

Figure 3.22 DM parasitic current loops during slow commutation process after switching-off of the transistor T_{b+}.

polarizing those diodes. Detailed topology of DM current loops that discharge $C_{MC+M,V-W}$ and charge $C_{MC+M,U-V}$ are presented in Figure 3.22.

According to Figure 3.22, after switching-off of transistor T_{b+} frequency converter's output current in phase V is reduced to zero, whereas current in motor winding of the same phase stays at similar level and is supported by parasitic DM current $I_{DM,Off}$ that flows in two subloops $I_{DM,Off1}$ and $I_{DM,Off2}$. Within the first DM current loop $I_{DM,Off1}$, parasitic capacitance $C_{MC+M,V-W}$ is discharged via two of motor winding, $V1 - V2$ and $W1 - W2$, by motor current that can be considered as approximately constant during entire switching-off time period.

The second DM current loop $I_{DM,Off2}$ forms a charging circuit of parasitic capacitance $C_{MC+M,U-V}$ that must be charged at the same time up to DC bus voltage V_{DC+}. That is why during the whole commutation process, the sum of voltages at capacitances $C_{MC+M,U-V}$ and $C_{MC+M,V-W}$ must stay at the same level, that is, V_{DC}, as they are continuously connected in series to DC bus voltage via switched-on transistors T_{c+} and T_{a-}. DM current loop $I_{DM,Off2}$ is formed by motor windings $V1 - V2$ and $W1 - W2$, switched-on transistors T_{c+} and T_{a-}, and DC bus capacitance. Finally, during the entire time period b of slow commutation process (Figure 3.21), while converters output current $I_{OUT,V} = 0$, total DM reloading current $I_{DM,Off}$ is a sum of subcurrents $I_{DM,Off1}$ and $I_{DM,Off2}$, discharging $C_{MC+M,V-W}$ and charging $C_{MC+M,U-V}$, respectively (3.20).

$$I_{DM,Off} = I_{DM,Off1} + I_{DM,Off2} \tag{3.20}$$

Simultaneously with DM parasitic current flowing during slow commutation process, CM parasitic currents also appear as a consequence of existing CM

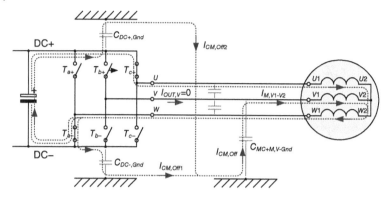

Figure 3.23 CM parasitic current loops during slow commutation process after switching-off the transistor T_{b+}.

parasitic capacitances. In total CM current $I_{CM,Off}$, two subcurrents $I_{CM,Off1}$ and $I_{CM,Off2}$ (3.21) can also be distinguished whose paths are presented in Figure 3.23.

$$I_{CM,Off} = I_{CM,Off1} + I_{CM,Off2} \tag{3.21}$$

One of them, $I_{CM,Off1}$, is forced only by electromagnetic force induced by current flowing previously in motor windings, whereas the second one $I_{CM,Off1}$ is additionally supported by DC voltage source that is the storage capacitance of DC bus. Both of them flow via parasitic capacitances of DC buses in relation to ground, $C_{DC+,Gnd}$ and $C_{DC-,Gnd}$, which are simultaneously reloaded according to the rule that the sum of their voltages should be equal to V_{DC} all the time, as they are continuously connected in series mutually and in parallel with DC bus.

CM currents generated during slow commutation process, as a part of motor current $I_{M,V1-V2}$, also reload CM parasitic capacitance of the commutated phase $C_{MC+M,V-Gnd}$, in the presented example from the initial potential of DC bus V_{DC+} up to final potential of V_{DC-}. After processes of reloading of all DM and CM parasitic capacitances are finished, which take place interdependently, voltage potential of converters terminal $V_{OUT,V}$ achieves level of V_{DC-} and allows it to switch-on forward current in the freewheeling diode integrated with the transistor T_{b-} (Figure 3.24).

As a consequence, motor current in commutated phase $I_{M,V1-V2}$, after short ringing transient (time period c, Figure 3.21) starts flowing again via converter's output V (time period d, Figure 3.21). The initial magnitude of the recovered output current $I_{OUT,V}$ is almost unchanged in relation to the value achieved

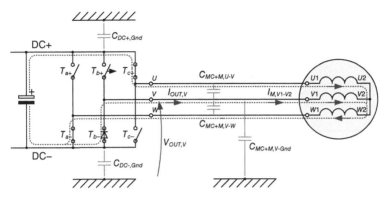

Figure 3.24 Distribution of converter's output currents after slow commutation process.

before commutation process; however, finished commutation output current $I_{OUT,V}$ begins to decrease according to the motor time constant.

Presented analysis of slow type of commutation processes allows estimating equivalent parasitic capacitance C_{par} of evaluated AC motor with feeding cable based on voltage switching time $t_{sw,slow}$ (time period *b*, Figure 3.21). Assuming that capacitances are reloaded with approximately constant motor current $I_{M,V1-V2}$, equivalent parasitic capacitance can be calculated using the formula (3.22), where V_{DC} is the switched voltage. An estimated calculation for the evaluated case, where switched voltage V_{DC} was about 600 V, commutated current $I_{M,V1-V2}$ about 3 A and recorded slow switching time 4 μs gives the value $C_{par} = 20\,\text{nF}$.

$$C_{par} = I_{M,V1-V2}\frac{t_{sw,slow}}{V_{DC}} \tag{3.22}$$

3.4.2 Current Transients at Inverter's Output

High-frequency components of CM and DM currents are generated simultaneously at each of three-phase terminals of output side of the converter during each voltage commutation from V_{DC-} to V_{DC+} and vice versa. Therefore, separation of CM and DM components requires individual measurements in all three phases simultaneously. An example of test setup used for more detailed measurement of parasitic HF current components is presented in Figure 3.25.

Measurement of converter's output currents in all phases and separately resultant CM current component using high-pass current probes inserted on the adequate wires allows verifying its summation. Recoded currents for fast commutation process are presented in Figure 3.26 and allow simplified analysis of CM and DM of parasitic current components distribution.

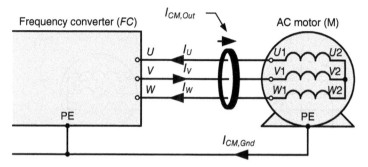

Figure 3.25 Measurement setup for HF current components at converter's output side.

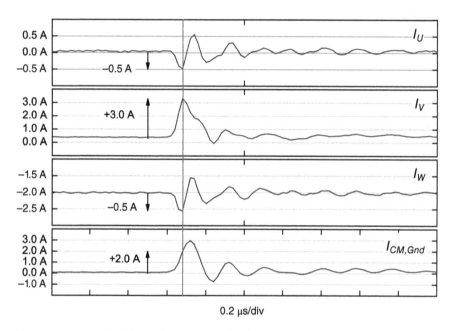

0.2 µs/div

Figure 3.26 Example of CM and DM currents distribution at converter's output side during fast type commutation.

Based on the theoretical investigation of CM and DM currents distribution presented in previous sections and the recorded data, it can be noticed that the maximum HF current component flows in the currently commutated phase. In the evaluated case, it is the current measured in phase V $I_{V,Out,HF}$, which is a sum of CM component $I_{CM,Out}$ and DM components $I_{U,Out,HF}$ and $I_{W,Out,HF}$ (3.23).

$$I_{V,Out,HF} = I_{U,Out,HF} + I_{W,Out,HF} + I_{CM,Out} \tag{3.23}$$

Distribution of DM (Figure 3.9) and CM (Figure 3.10) parasitic capacitances of motor cable and motor windings causes that part of HF transient current in the commutated phase $I_{V,Out,HF}$ to return back to the converter's DC bus as DM components. Assuming that during commutation the phase U is connected to V_{DC+} and the phase W to V_{DC-} via switched-on transistors, two DM current subcomponents flow in these two phases, $I_{U,Out,HF}$ and $I_{W,Out,HF}$. These two DM current components are usually of similar magnitude. If slight difference exists, it is related to converter's load HF impedance asymmetry.

The above discussion about current directions can be analyzed only at a very initial period of the commutation process. After achieving local maximum values of phase currents, marked in Figure 3.26 by vertical line, directions of currents are difficult to determine because the following ringing oscillations of DM and CM components have usually slightly different frequencies and phase shifts. Finally, we can obtain formulas for determining CM current as (3.24), which can also be measured directly and for differential currents in each phases as (3.25)–(3.27).

$$I_{CM,Out} = I_{V,Out,HF} - I_{U,Out,HF} - I_{W,Out,HF} \tag{3.24}$$

$$I_{DM,V,HF} = I_{V,Out} - I_{CM,Out} \tag{3.25}$$

$$I_{DM,U,HF} = I_{U,Out,HF} = \frac{1}{2}(I_{V,Out,HF} - I_{CM,Out}) \tag{3.26}$$

$$I_{DM,W,HF} = I_{W,Out,HF} = \frac{1}{2}(I_{V,Out,HF} - I_{CM,Out}) \tag{3.27}$$

Summarizing, the maximum HF parasitic current that flows in the commutated phase is a sum of CM current and DM currents in the other two phases. From time domain perspective, the HF current accompanying the DC bus voltage commutation exiting from commutated phase returns back to the inverter as two DM current components flowing in the other two phases and CM component that returns back through motor-converter ground connections.

The above investigation of HF parasitic currents distribution into CM and DM modes has been done for relatively low amplitudes of LF component of output currents to avoid saturation of high-pass current probes. The results obtained are particularly important and useful for further analysis because in high-power ASD applications where LF components of output current are high, the measurement of HF components of currents in each separate phase is significantly more difficult.

The measurement difficulty of parasitic HF current components generated at output side of frequency converter is related to simultaneous existence of high magnitudes of LF current components of frequency relatively close to power frequency and very short, usually well below 1 μs, current transients of significant magnitudes as well. Such measuring conditions require the use of high-pass current probes of high sensitivity and linearity within HF band that simultaneously tolerate high levels of LF magnetic field accompanying the high magnitudes of AC motor operating currents.

3.5 Summary

Conducted emission of ASD is closely related to HF currents and voltage transient generation at output side of FC due to fast voltage switching and unavoidable parasitic capacitances distributed along motor feeding cable and motor windings. Final levels of generated conducted emission and its distribution throughout ASD are entirely related to CM and DM current paths formed mostly unintentionally by parasitic capacitances of adequate types, CM and DM, which are inseparably linked to each other. Simultaneous existence of CM and DM components of HF parasitic currents is one of the major difficulties of their identification, analysis, and limitation.

Generated high-voltage steepness at converter's output-side terminals during transistor's commutation is the initial and main cause of HF current transient generation; therefore, very common investigation of only CM voltage issues is a significant simplification that may lead to inconsistencies. Essentially, resultant voltage steepness generated at converter's output terminals is an effect of value of switched DC voltage and switching time, which is a key parameter of a transistor. However, the presented analysis reveals that switching time for commutation after which magnitude of output current is going to decrease is commonly much longer and essentially depends on parasitic capacitances of converters load circuitry and magnitude of the commutated current.

The differences in HF currents transient generated during switching between active and passive output vectors are not that significant; therefore, PWM switching pattern modifications related to increased use of passive vectors will not significantly effect conducted emission. HF current transients generated at converter's output terminals return back to the DC bus circuit as DM and CM currents. Unfortunately, due to typical CM and DM parasitic capacitances distribution, CM return current is usually greater and more significant from conducted emission point of view, because it can return to the converter through diverse grounding paths.

References

1 K. Lee and G. Nojima, "Quantitative power quality and characteristic analysis of multilevel pulsewidth-modulation methods for three-level neutral-point-clamped medium-voltage industrial drives," *IEEE Transactions on Industry Applications*, vol. 48, no. 4, pp. 1364–1373, July 2012.

2 Malinowski, M. P. Kazmierkowski, and A. M. Trzynadlowski, "A comparative study of control techniques for PWM rectifiers in AC adjustable speed drives," in *The 27th Annual Conference of the IEEE Industrial Electronics Society, 2001 (IECON'01)*, vol. 2, 2001, pp. 1114–1118. Available at http://ieeexplore.ieee.org/stamp/stamp.jsp?arnumber=975936

3 H. Soltani, P. Davari, F. Zare, and F. Blaabjerg, "Effects of modulation techniques on the input current interharmonics of adjustable speed drives," *IEEE Transactions on Industrial Electronics*, vol. 65, no. 99, pp. 1–1, 2017.

4 J. Adabi, A. Boora, F. Zare, A. Nami, A. Ghosh, and F. Blaabjerg, "Common-mode voltage reduction in a motor drive system with a power factor correction, " *IET Power Electronics*, vol. 5, no. 3, pp. 366–375, March 2012.

5 D. Han, C. T. Morris, and B. Sarlioglu, "Common-mode voltage cancellation in PWM motor drives with balanced inverter topology," *IEEE Transactions on Industrial Electronics*, vol. 64, no. 4, pp. 2683–2688, April 2017.

6 C. Jettanasen, "Reduction of common-mode voltage generated by voltage-source inverter using proper PWM strategy," in *2012 Asia-Pacific Symposium on Electromagnetic Compatibility*, May 2012, pp. 297–300.

7 D. Jiang, F. Wang, and J. Xue, "PWM impact on CM noise and ac CM choke for variable-speed motor drives," *IEEE Transactions on Industry Applications*, vol. 49, no. 2, pp. 963–972, March 2013.

8 A. Videt, P. L. Moigne, N. Idir, P. Baudesson, J. J. Franchaud, and J. Ecrabey, "Motor overvoltage limitation by means of a new EMI-reducing PWM strategy for three-level inverters," *IEEE Transactions on Industry Applications*, vol. 45, no. 5, pp. 1678–1687, Sept. 2009.

9 Y. Zhao, G. Tan, L. Zhang, and Y. Zhang, "Suppression of electromagnetic interference (EMI) in PWM voltage source rectifier (VSR) based on chaotic SVPWM," in *2011 IEEE Power Engineering and Automation Conference*, vol. 2, Sept. 2011, pp. 99–102.

10 Palma and P. Enjeti, "An inverter output filter to mitigate dv/dt effects in PWM drive system," in *Seventeenth Annual IEEE Applied Power Electronics Conference and Exposition, 2002 (APEC 2002)*, vol. 1, 2002, pp. 550–556. Available at http://ieeexplore.ieee.org/stamp/stamp.jsp?arnumber=989298

11 G. L. Skibinski, R. M. Tallam, M. Pande, R. J. Kerkman, and D. W. Schlegel, "System design of adjustable speed drives, part 2: System simulation and ac line interactions," *IEEE Industry Applications Magazine*, vol. 18, no. 4, pp. 61–74, July 2012.

12 P. Musznicki, *Conducted EMI Identification in Power Electronic Converters: Modeling of EMI Generation and Propagation Using Circuit Simulation and Wiener Filtering Methods*, VDM Publishing, 2009.

13 N. Oswald, B. H. Stark, N. McNeill, and D. Holliday, "High-bandwidth, high-fidelity in-circuit measurement of power electronic switching waveforms for EMI generation analysis," in *2011 IEEE Energy Conversion Congress and Exposition*, Sept. 2011, pp. 3886–3893.

14 W. Chen, P. Chen, L. Feng, H. Chen, and Z. Qian, "Frequency model for differential-mode conducted EMI prediction in three-phase PWM inverter," in *Fortieth IAS Annual Meeting Industry Applications Conference, 2005. . Conference Record of the 2005*, vol. 4, Oct. 2005, pp. 2816–2819.

15 G. Ala, G. C. Giaconia, G. Giglia, M. C. D. Piazza, and G. Vitale, "Design and performance evaluation of a high power-density EMI filter for PWM inverter-fed induction-motor drives," *IEEE Transactions on Industry Applications*, vol. 52, no. 3, pp. 2397–2404, May 2016.

16 B. Revol, J. Roudet, J. L. Schanen, and P. Loizelet, "EMI study of three-phase inverter-fed motor drives," *IEEE Transactions on Industry Applications*, vol. 47, no. 1, pp. 223–231, Jan. 2011.

17 K. Chen, Z. Jin, and H. Chen, "Effect of common-mode interference on communication performance of a motor drive system," in *2016 IEEE Vehicle Power and Propulsion Conference (VPPC)*, Oct. 2016, pp. 1–6.

18 M. Jin, M. WeiMing, P. Qijun, K. Jun, Z. Lei, and Z. Zhihua, "Identification of essential coupling path models for conducted EMI prediction in switching power converters," *IEEE Transactions on Power Electronics*, vol. 21, no. 6, pp. 1795–1803, Nov. 2006.

19 H. Cheaito, M. S. Diop, M. Ali, E. Clavel, C. Vollaire, and L. Mutel, "Virtual bulk current injection: modeling EUT for several setups and quantification of CM-to-DM conversion," *IEEE Transactions on Electromagnetic Compatibility*, vol. 59, no. 3, pp. 835–844, June 2017.

20 A. Kempski and R. Smolenski, "Decomposition of EMI noise into common and differential modes in PWM inverter drive system," *Electrical Power Quality and Utilisation, Journal*, vol. 12, no. 1, pp. 53–58, 2006.

21 I. F. Kovacevic, T. Friedli, A. M. Musing, and J. W. Kolar, "3-D electromagnetic modeling of parasitics and mutual coupling in EMI filters," *IEEE Transactions on Power Electronics*, vol. 29, no. 1, pp. 135–149, 2014. Available at http://ieeexplore.ieee.org/stamp/stamp.jsp?arnumber=6484987

22 A. Axelrod, "Experimental study of DM-to-CM and vice-versa conversion effects in balanced signal and power line filters," in *2003 IEEE International Symposium on Electromagnetic Compatibility(EMC '03). , vol. 1, 2003, pp. 599–602. Available at http://ieeexplore.ieee.org/stamp/stamp.jsp?arnumber=1428330

23 I. F. Kovaeevi, T. Friedli, A. M. M using, and J. W. Kolar, "Full PEEC modeling of EMI filter inductors in the frequency domain," *IEEE Transactions on Magnetics*, vol. 49, no. 10, pp. 5248–5256, Oct. 2013.

24 M. Kamikura, Y. Murata, and A. Nishizawa, "Investigation on the mode conversion between common-mode and differential-mode noises in EMI filters for power electronics circuits," in *2013 International Symposium on Electromagnetic Compatibility (EMC EUROPE)*, 2013, pp. 557–560. Available at http://ieeexplore.ieee.org/stamp/stamp.jsp?arnumber=6653365

25 S. Wang and F. C. Lee, "Investigation of the transformation between differential-mode and common-mode noises in an EMI filter due to unbalance," *IEEE Transactions on Electromagnetic Compatability*, vol. 52, no. 3, pp. 578–587, 2010. Available http://ieeexplore.ieee.org/stamp/stamp.jsp?arnumber=5406077

26 M. Heldwein, J. Biela, H. Ertl, T. Nussbaumer, and J. Kolar, "Novel three-phase CM/DM conducted emission separator," *IEEE Transactions on Industrial Electronics*, vol. 56, no. 9, pp. 3693–3703, 2009.

27 J. Stahl, D. Kuebrich, A. Bucher, and T. Duerbaum, "Characterization of a modified LISN for effective separated measurements of common mode and differential mode EMI noise," in *Energy Conversion Congress and Exposition (ECCE)*, IEEE, 2010, pp. 935–941. Available at http://ieeexplore.ieee.org/stamp/stamp.jsp?arnumber=5617888

28 L. Zhang, K. Wang, J. Meng, and W. Ma, "The fabrication and application of CM/DM interference separation network based on transmission-line transformer," in *2010 Asia-Pacific Symposium on Electromagnetic Compatibility (APEMC)*, 2010, pp. 1618–1621.

29 S. W. Pasko, M. K. Kazimierczuk, and B. Grzesik, "Self-capacitance of coupled toroidal inductors for EMI filters," *IEEE Transactions on Electromagnetic Compatibility*, vol. 57, no. 2, pp. 216–223, April 2015.

30 S. Braun, "A novel time-domain EMI measurement system for measurement and evaluation of discontinuous disturbance according to CISPR 14 and CISPR 16," in *2011 IEEE International Symposium on Electromagnetic Compatibility (EMC)*, pp. 480–483. Available at http://ieeexplore.ieee.org/stamp/stamp.jsp?arnumber=6038359

31 Y.-S. Lee, Y.-L. Liang, and M.-W. Cheng, "Time domain measurement system for conducted EMI and CM/DM noise signal separation," in *International Conference on Power Electronics and Drives Systems, 2005 (PEDS'05)*, vol. 2, 2005, pp. 1640–1645.

32 L. Ran, S. Gokani, J. Clare, K. J. Bradley, and C. Christopoulos, "Conducted electromagnetic emissions in induction motor drive systems. I. Time domain analysis and identification of dominant modes," *IEEE Transactions on Power Electronics*, vol. 13, no. 4, pp. 757–767, 1998. Available at http://ieeexplore.ieee.org/xpls/abs_all.jsp?arnumber=704152

33 J. Xue, F. Wang, and B. Guo, "EMI noise mode transformation due to propagation path unbalance in three-phase motor drive system and its implication to EMI filter design," in *2014 IEEE Applied Power Electronics Conference and Exposition (APEC 2014)*, March 2014, pp. 806–811.

34 Z. Du, L. M. Tolbert, and J. N. Chiasson, "Modulation extension control for multilevel converters using triplen harmonic injection with low switching frequency," in *Twentieth Annual IEEE Applied Power Electronics Conference and Exposition, 2005 (APEC'05)*, vol. 1, March 2005, pp. 419–423.

35 W. Jiang, W. Ma, J. Wang, W. Wang, X. Zhang, and L. Wang, "Suppression of zero sequence circulating current for parallel three-phase grid-connected converters using hybrid modulation strategy," *IEEE Transactions on Industrial Electronics*, vol. 65, no. 99, pp. 1, 2017.

4

Propagation of Motor-Side-Originated Conducted Emission Toward the Power Grid

You can't stop the waves, but you can learn to surf.
Jon Kabat-Zinn

4.1 Characteristic External Loops of Common Mode Currents in the ASD

Common mode currents generated at the motor side of frequency converter during each commutation process of output voltage are circulating in a loop consisting of motor cable, motor winding, grounding connections between FC and motor, and internal components of FC (Figure 4.1). The spectral content of those circulating HF currents strongly depends on slope rates of voltage transients generated at converters outputs and distribution of parasitic capacitances distributed along all parts of the motor side of ASD. There are many known reports of research that document that the HF current transients generated at motor side of FC propagate toward the power grid and therefore can significantly influence overall conducted emission of the ASD [1–3].

Assuming that conducted emission levels of the FC output side are known or possible to determine, estimation of conducted emission levels at the grid side of the ASD requires to recognize the output-to-input transfer characteristic. Determination of this characteristic requires analysis of FC internal coupling paths that influence frequency-dependent coupling factors. Comprehensive knowledge of HF coupling phenomena is also helpful for efficient design of internal filtering components used for decreasing conducted emission leakage from the motor side to the grid side of FC.

Based on the already known results of research, there are two primary types of possible couplings presented in Figure 4.1: by the FC external grounding

High Frequency Conducted Emission in AC Motor Drives Fed by Frequency Converters: Sources and Propagation Paths, First Edition. Jaroslaw Luszcz.

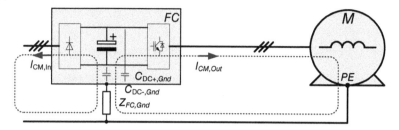

Figure 4.1 Internal and external coupling paths between motor side and grid side of frequency converter.

impedance $Z_{FC,Gnd}$ and by parasitic capacitances of internal components of FC, especially between DC buses C_{DC+}, Gnd, C_{DC-}, Gnd, and the grounded chassis [4–6]. Therefore, low impedance of grounding connection within wide frequency range of FC is strongly recommended and successfully used for decreasing of resulting emission of ASD. Intended modification of parasitic coupling inside FC—to achieve conducted emission decrease—is significantly more difficult and usually requires advanced analysis considering not only emission limitation but also entire functionality of FC.

4.2 Common Mode Currents Coupling Paths Inside Frequency Inverter

Many research reports recognize parasitic capacitances between power transistors semiconductor substrate and usually grounded heat sinks as the main coupling paths of conducted emission propagation inside the FC [7–9]. Taking into account that in power transistors the most significant parasitic capacitance for CM currents propagation is the collector-to-heat sinks capacitance $C_{IGBT,Gnd}$ [10,11], in commonly used topology of the output bridge those capacitances can be specifically assigned to underline their influence on CM currents flow. Collector terminals of upper transistors T_{a+}, T_{b+}, T_{c+} are galvanically connected to the positive DC bus V_{DC+}; therefore, collector-to-ground parasitic capacitances of these transistors increase the overall capacitance of positive DC bus in relation to ground $C_{DC,Gnd}$ (Figure 4.2).

Similarly, parasitic collector-to-heat sink capacitances of lower transistors T_{a-}, T_{b-}, T_{c-} can be assigned to parasitic capacitances of each of the inverters output terminals U, V, and W, thus added to motor cable wire-to-ground capacitances $C_{W,Gnd}$, because they are connected in parallel. Therefore, the impact of transistor-to-ground capacitances $C_{IGBT,Gnd}$ on conducted disturbances propagation can be analyzed cumulatively together with equivalent

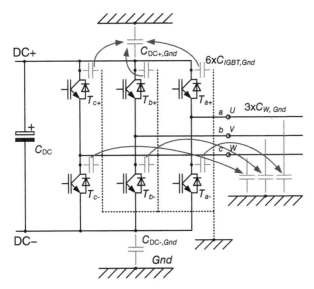

Figure 4.2 Transistor-to-ground capacitances distribution in relation to converter's output terminals and DC bus.

parasitic capacitances of DC bus and the converter's output terminals in relation to ground.

Significance of contribution of parasitic capacitances of upper transistor's T_{a+}, T_{b+}, T_{c+} to the total equivalent parasitic capacitance between the positive DC bus and ground can vary depending on the rated power of the ASD and coveter's constructional details. Regarding parasitic capacitances of the lower transistor's T_{a-}, T_{b-}, T_{c-}, they are usually much smaller than the sum of parasitic CM capacitances of the motor cable and windings for the majority of ASD applications, even with a short motor cable. Nevertheless, their influence is significant because they are localized very close to the output terminals with regard to corresponding capacitances distributed along the motor cable and windings. Summarizing, for analysis of CM currents propagation in ASDs, parasitic capacitances of the output inverter's transistors, in relation to grounded heat sink, can be considered as combined together with parasitic capacitances of the positive DC bus and of the output terminals.

4.3 Transfer of Common Mode Currents via DC Bus

Analysis of propagation of conducted emission generated at the output side of FC toward the diode rectifier requires parasitic parameters (inductances and capacitances) of DC bus bar connections and possible filtering components,

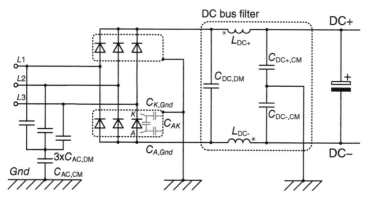

Figure 4.3 Simplified representation of CM current couplings in DC bus.

which can be inserted between the DC bus capacitor bank and the diode rectifier output. CM and DM parameters of the DC bus bar connection, regardless of whether they are real or parasitic, can be represented in a simplified, unified form as a circuit model, with its topology similar to one of those commonly used for EMI filtering in power lines. In majority of applications, the DC bus filter configuration is comprised of serial inductances L_{DC+} and L_{DC-}, DM capacitance $C_{DC,DM}$, and CM capacitances $C_{DC+,CM}$ and $C_{DC-,CM}$ (Figure 4.3) and is sufficient to adequately represent the most essential effects.

In a typical FC used in a low-power ASD, DC chokes are often used to decrease LF harmonics emission of the three-phase full-wave rectifier into the power grid. Inductances of these chokes are usually significantly higher than parasitic inductances of DC bus connection. The secondary effect of the impedance of DC choke implemented in DC bus, well designed especially in HF range, can have significant impact on transfer of conducted emission generated at the motor side of converter to the grid side. Nevertheless, one of the implementation problems is that in such DC chokes with a relatively high inductance and a high rated DC current, it is usually difficult to minimize parasitic capacitances of windings that can significantly reduce its potential benefits for conducted emission reduction.

On the other hand, DC chokes, even if they are not specially designed with consideration of their performance in HF range, usually influence CM currents transfer significantly, not always positively, and thus have to be considered in the analysis. DC chokes with coupled windings, generally recommended for conducted emission reduction, are much less commonly used in the DC bus of FC, even though they can be more effective, because they allow achieving higher level of balance between positive and negative buses of the DC link connection.

Parasitic DM capacitance $C_{DC,DM}$ of the DC bus has usually significant value that is quite often additionally increased by intentionally implemented small capacitor with low serial inductance. The main purpose of this capacitor is better attenuation of high-frequency DM voltage ripples in the DC bus. This capacitance also influences CM currents indirectly by changing the ratio between CM an DM capacitances.

CM capacitors $C_{DC+,CM}$ and $C_{DC-,CM}$ can be implemented into the DC bus—primarily for maintaining high-frequency CM currents distribution inside the FC [12,13]. Application of those capacitances allows decreasing the transfer of CM currents originated as the effect of PWM modulation at the motor side of the inverter, and also can be used for improvement of balance of parasitic capacitances of the DC+ and DC− buses that usually have different values primarily [14,15]. Sizing of CM-type capacitances inserted in the DC bus is problematic and has to be carefully optimized because these capacitances also cause negative effects. The main disadvantages are the increase of the motor-side CM currents and general increase of the overall residual current of ASD, which is limited by safety standards [16,17]. Therefore, their effective use requires coordination of parameters of the motor and the motor cable. If they are not properly sized, this can lead to their ineffectiveness or even drawbacks.

4.4 Transfer of CM Currents Through Diode Rectifier

The grid side diode rectifier in ASD in a high-frequency range exhibits three types of parasitic capacitances: anode–cathode C_{AK}, anode-to-ground $C_{A,Gnd}$, and cathode-to-ground $C_{A,Gnd}$ (Figure 4.3). In practice, parasitic couplings between rectifier diodes and ground are mainly associated with relatively higher parasitic capacitances between diode substrates and grounded heat sink of the rectifier module.

Similar to the output inverter, capacitances of rectifier diodes to ground can be lumped to equivalent CM capacitances $C_{DC+,CM}$, $C_{DC-,CM}$ of the positive and negative DC buses and input AC terminals $C_{AC,CM}$. Therefore, these capacitances do not have to be analyzed separately. The remaining diode parasitic self-capacitance C_{AK} determined for nonconducting state can be find in a typical parameter list of diodes; nevertheless, this capacitance is not significantly high in relation to diode-to-heat sink capacitances.

Furthermore, during typical operation of loaded three-phase full-wave rectifier, each time at least two diodes are in a conducting state that simply means short circuit also for HF currents. HF currents transfer dominates in the currently conducting rectifier branches, and thus transfer through nonconducting diodes can be neglected.

The phenomenon of transfer of high-frequency CM currents through loaded diode rectifier can be easily explained in a clear way by analysis of grid-side CM currents of the ASD during low load operation where the DC output current

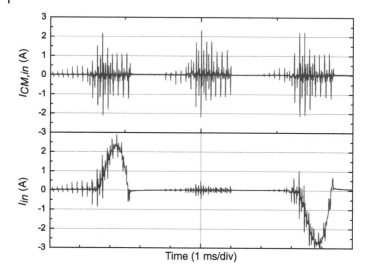

Figure 4.4 Correlation of a phase input current and the input CM current of FC for low load of the ASD with no continuous DC current in the rectifier output.

of the rectifier is not continuous. The correlation between input current I_{in} in one of the phases and the input CM current $I_{CM,in}$ is presented in Figure 4.4.

This comparison clearly shows that transfer of CM currents originated from PWM modulation of the inverter output toward the grid side is much higher within periods of time when the rectifier input current is nonzero. In the observed phase, after each decrease of rectifier input current to zero, the input CM current decreases substantially; so its influence on the overall emission level can be omitted. In the central part of the presented waveforms where the measured phase current I_{in} is close to zero, the observed increase of CM current is correlated to the input current conducted in other phase than currently monitored, since the input CM current $I_{CM,in}$ has been measured in all three-phase wires simultaneously.

4.5 DC Bus CM Voltage Ripples as Means of Transfer of Conducted Emission

The DC bus connection in ASD is usually being optimized for providing well-filtered DC voltage with possibly lowest level of AC ripples and lowest serial inductance within wide frequency range. A high-quality DC voltage source is desirable for obtaining high performance of output frequency inverter operation with minimized exposure to overvoltage of a switching transistor during output voltage commutations. From the conducted emission point of view, the

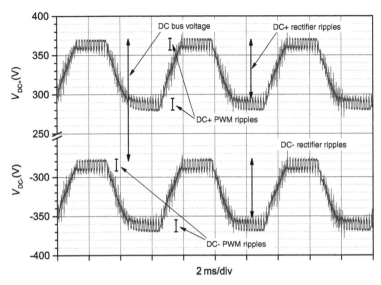

Figure 4.5   DC bus voltages: positive bus bar V_{DC+}—upper waveform; negative bus bar V_{DC-}—lower waveform.

DC bus is an essential part of the ASD that interconnects the grid side with the output side, with high capability of efficient power flow.

The DC bus connection between the input rectifier bridge and the output transistor bridge, designed to achieve high-performance filtering of LF, grid-related voltage harmonics, and low ripples at DC bus output, can also be considered as a natural filter for HF conducted emission. Unfortunately, parasitic capacitances of DC bus connections designed primarily for high voltage and current capacity can significantly reduce filtering effectiveness, especially within higher frequency bands of conducted emission [15].

Observation of DC bus voltages—particularly referenced to ground—is rarely undertaken in EMC analysis. Study of DC bus voltage waveforms is challenging because of severe difficulties with reliable measurement and selection of HF components that are usually significantly lower than dominating LF and DC components. Nevertheless, analysis of DC bus voltages in ASD can be a very important source of evidence useful for better understanding of conducted emission propagation inside FC. Examples of DC+ and DC− CM voltages recorded in the evaluated ASD are presented in Figure 4.5.

When analyzing recorded waveforms, it can be noticed that voltage changes at positive and negative buses are synchronous; each increase of voltage at positive bus corresponds to decrease of voltage at negative bus. There are two dominating, significantly different from one another, types of voltage variations visible: one correlated to the three-phase full-wave grid rectifier with exposed dominating frequency three times higher than the power frequency (300 Hz), and the second one related to the PWM carrier frequency (6 kHz). Based on the presented waveforms, CM and DM components of the DC bus voltage can be determined using Eqs. (4.1) and (4.2):

$$V_{DC,DM} = V_{DC+} - V_{DC-} \qquad (4.1)$$

$$V_{DC,CM} = \frac{V_{DC+} + V_{DC-}}{2} \qquad (4.2)$$

Decomposition of DC bus voltages into DM and CM components shows that the CM voltage component of the DC bus contains voltage ripples of significantly higher magnitudes in relation to DM. Voltage ripples occurring at the DC bus as a mid part of the FC can propagate to other parts that are coupled by parasitic capacitances or by capacitors introduced by purpose. Values of parasitic capacitances are typically in a range starting from nanofarad and depend on the DC bus connections dimensions that are strongly related to the rated power of the converter. Such capacitances are high enough to efficiently transfer power of signals of frequency of the PWM carrier, which in typical low and medium power ASDs are in a range starting from few kilohertz.

Referring to Figures 4.2 and 4.3, many of parasitic capacitances of the grid rectifier and the output inverter can be integrated to equivalent capacitances referenced to the input and output terminals of the FC. Only DC bus capacitances that are repeatedly reconnected to the input and output terminals have to be analyzed individually. Repetitive switchings and reloading of capacitances localized between DC buses and the input and output terminals of the FC are essential in analysis of conducted emission propagation. Therefore, the DC bus, from the conducted emission point of view, is a central and commonplace of the FC where input- and output-side CM currents are coupled [15,18]. Modification and optimization of intentional and parasitic inductances and capacitances of the DC bus allows influencing significantly conducted emission propagation, thus possibly to reduce requirements for additional, external filtering.

4.5.1 Diode Rectifier-Related CM Voltage Ripples

In three-phase full-wave rectifiers, voltages at both output terminals DC+ and DC− float in relation to ground potential due to diodes commutation. The maximum voltage ripples observed at the rectifier output loaded by resistance can achieve the level of $\sqrt{2}/2 * V_{L-N,RMS}$. In ASDs, relatively large DC bus capacitances are usually connected to rectifier outputs via DC chokes that are used for limitation of low-order harmonics emission into the power grid and decreasing of magnitudes of input current peaks.

Limitation of differential ripples of voltage at the DC bus due to the fundamental requirements of DC powering of frequency inverter also decrease CM voltage ripples at the DC bus, which are essential for conducted emission generation. The level of output voltage ripples reduction depends primarily on parameters of the DC bus LC filter but also on rectifier's load; therefore, it can change during typical operation of an ASD with varying load condition. Typically in ASD, DM voltage ripples at DC bus are usually within the range of 10–30% of DC components of the output voltage for full load condition.

Fundamental frequency of voltage ripples at the rectifier output is three times higher than the power grid frequency, which is an effect of six pulse rectifying bridge. This low-frequency harmonic emission phenomena are usually classified as power quality issue rather than conducted emission issue by the EMC community.

Nevertheless, harmonics of voltages of such frequency $3 * f_{PG}$, and its multiples, can result in significant increase of stray currents flowing via parasitic capacitances and filter capacitors. Increased leakage currents to ground, which is usually also a protective earth (PE), can cause severe problems with multifunctional operation of residual current devices (RDC) widely used for shock protection. Waveforms of voltage ripples are as well evidently nonsinusoidal (Figure 4.5); therefore, higher odd harmonic components are present and they are primarily of triplen order from the perspective of the power system H9, H15, H21, and so on. Triplen harmonics are called zero-sequence harmonics that can have several negative effects for the power system, for example, current flow in neutral wire because they are in phase with each other.

4.5.2 Frequency Inverter-Related CM Voltage Ripples

PWM-related voltage ripples are superimposed on rectifier-related ones and their occurrence is closely correlated with FC output voltage switching according to PWM pattern. Therefore, they occur regularly according to the PWM carrier frequency and, on the other hand, irregularly because of continuous PWM duty cycle changes.

Two categories of the PWM-related DC voltage ripple shapes can be noticed on the waveforms presented in Figure 4.6. First category, visible on the right-hand side of the presented waveform in the time interval approximately between 8 and 10 ms, is more rectangular-like in time subperiods within which DC bus voltage is relatively constant, not changing due to charging–discharging process of DC bus capacitors bank. Second category, with more transient shapes, in time subperiods (Figure 4.6 approximately between 6 and 8 ms) where DC bus voltage is changing due to charging–discharging process of the DC bus main capacitor bank.

Magnitudes of the PWM-related voltage ripples depend mainly on parasitic parameters of positive and negative DC buses and the rectifier load current. In

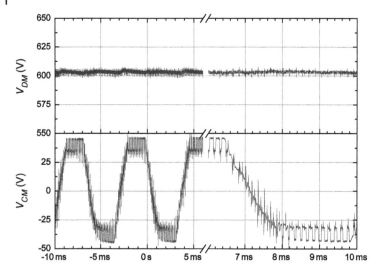

Figure 4.6 Calculated DC bus voltage components: differential mode $V_{DC,DM}$—upper waveform; common mode $V_{DC,CM}$—lower waveform.

ASDs, inverter-related voltage ripples usually reach values of a few percent of differential voltage of DC bus V_{DC}; it is significantly less than magnitudes of the rectifier's originated low-frequency ripples. Since their frequencies are usually much higher than low-order harmonics of power frequency, their influence on conducted emission of ASDs can be essential. Generally CM voltage $V_{DC,CM}$ occurring at the DC bus connection in frequency converter have fundamental impact on transfer of HF currents transients generated at the output side of the frequency inverter toward rectifier side of the converter.

4.6 Transfer of CM Currents from Motor Side to Grid Side of Frequency Converter

PWM-related CM currents transfer from motor side of the FC to the power grid can be explained using simplified circuit model that takes into account the coupling capacitances discussed in previous sections. The main loops of circulating CM current are presented in Figure 4.7 where the most significant capacitive couplings to ground are represented by lumped capacitances connected to external input terminals $L1$ and $L2$—$C_{AC-In,CM}$, output terminal V—$C_{MC,W-Gnd}$, and positive and negative DC buses $C_{DC-,Gnd}$, $C_{DC+,Gnd}$.

One of the most commonly occurring output voltage switching patterns starts from disconnection of one of the FC output terminals from one DC bus and after finishing commutation process connection of the same terminal to

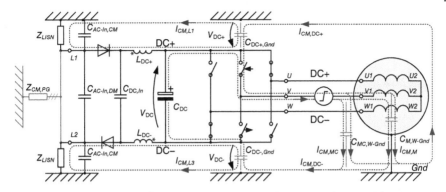

Figure 4.7 Coupling paths of CM currents generated at the motor side toward the power grid.

the opposite DC bus. In the presented case, output terminal V is disconnected from the DC− bus and afterward connected to the DC+ bus.

Assuming that before described disconnection, the output current in the commutated phase was positive (flowing from terminal V), after reconnection of this terminal to the DC+ bus, there will be an increase of magnitude of the output current in this phase (V), which means that fast switching pattern will be generated, according to the explanation presented in Figure 3.27.

In case the output current was negative before commutation process, after reconnection a decrease of output current will take place. In this case the switching process will be much slower and depend on the freewheeling diode conduction time that is related to magnitude of commutated current. Slower commutation process results in generation of lower CM currents, which means that from the point of view of conducted emission, the fast switching patterns are the most influential.

After fast commutation patterns for which switching time depends mainly on transistor dynamic characteristics, the output parasitic capacitance of the switched wire of motor feeding cable $C_{MC,W-Gnd}$ and switched phase of motor windings $C_{M,W-Gnd}$, which was previously charged to voltage level of DC− bus, starts to be reloaded up to the opposite voltage of DC+ bus. This reloading process induces CM current $I_{CM,DC-}$ that flows through the parasitic capacitance of negative DC bus to ground $C_{DC-,Gnd}$ as well. Therefore, the absolute voltage at $C_{DC-,Gnd}$ in relation to ground will decrease. At the same time, because the sum of voltages V_{DC+} and V_{DC-} at DC bus parasitic capacitances $C_{DC+,Gnd}$ and $C_{DC-,Gnd}$ have to be equal to DC bus voltage V_{DC}, simultaneously accompanying second CM current $I_{CM,DC-}$ is enforced, which increases the voltage at $C_{DC+,Gnd}$ to fulfill Eq. 4.3.

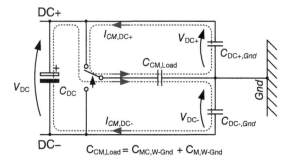

Figure 4.8 Simplified representation of output-side CM current loops with exposed midpoint of DC bus voltage V_{DC} parasitically connected to ground via capacitances $C_{DC+,Gnd}$ and $C_{DC-,Gnd}$.

$$V_{DC+} + V_{DC-} = V_{DC} \tag{4.3}$$

Currents $I_{CM,DC+}$ and $I_{CM,DC-}$ are summing up in the switched wire of load and in ground connections that are returning paths for these currents back to DC bus. The parasitic capacitances of DC bus in relation to ground $C_{DC+,Gnd}$ and $C_{DC-,Gnd}$ are the essential parts of output-side CM current loops that cause DC+ and DC− potentials floating in relation to ground due to transistor commutation processes. Simplified equivalent circuit that shows more clearly how DC bus is changing their voltages V_{DC+} and V_{DC-} in relation to ground during charging and discharging of parasitic capacitance $C_{DC+,Gnd}$ and $C_{DC-,Gnd}$ is presented in Figure 4.8.

Levels of DC bus CM voltage fluctuations due to PWM switching processes can be estimated based on ratio between equivalent parasitic capacitances of converter's load $C_{CM,Load}$ (4.4) and DC bus $C_{DC+,Gnd}$ and $C_{DC-,Gnd}$ (4.5):

$$C_{CM,Load} = C_{MC,W-Gnd} + C_{M,W-Gnd} \tag{4.4}$$

$$\Delta V_{DC+} = \Delta V_{DC-} = \frac{2C_{CM,Load}}{C_{DC+,Gnd} + C_{DC-,Gnd}} V_{DC} \tag{4.5}$$

Parasitic equivalent capacitance of load $C_{CM,Load}$ is a sum of parasitic capacitances of the motor cable wire and the motor windings determined for frequency close to the PWM carrier frequency. Therefore, DC bus CM voltage ripples, which take place after each output inverter switching, depend on the FC load frequency response impedance, in particular the motor feeding cable CM capacitance as a most varying parameter of the ASD. In practice, it is dif-

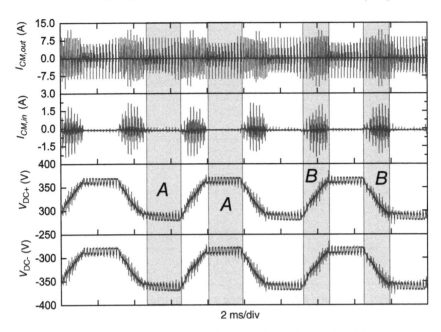

2 ms/div

Figure 4.9 Input and output CM currents of converter's correlated with DC bus voltages referenced to ground.

ficult to determine values of parasitic capacitances of load because the lengths of the motor cable used in a particular application can differ significantly. In Figure 4.9, CM currents at the input and output of the FC recorded in time domain with correlation to positive and negative DC bus voltages ripples are presented.

Based on this observation, it can be seen that PWM-related CM currents generated continuously at the output side are transferred toward the grid side evidently unevenly, only within certain time intervals. Those intervals are closely correlated to the rectifier's diodes conduction and blocking states. During diodes conduction intervals (marked as A), while DC bus voltages V_{DC+} and V_{DC-} are changing due to the loading process of DC bus capacitance C_{DC}, the observed grid-side CM current $I_{CM,In}$ is significantly higher. In between regularly repeated intervals of type *A*, where DC bus voltages V_{DC+} and V_{DC-} are not changing (marked as *B*) and all diodes are in nonconducting state, CM input current $I_{CM,In}$ is much lower. The presented waveform was intentionally recorded for unloaded motor with noncontinuous output current of the rectifier for better presentation of the process of HF current transients transfer via diode rectifier from DC bus to the grid side. In frequency considerably loaded converters, continuous current at rectifier's DC output is usually expected; therefore, CM current transfer is more regular and effective.

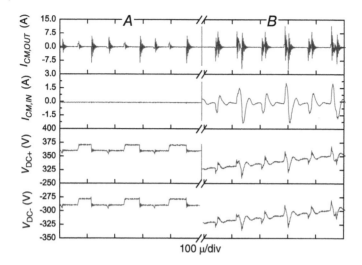

100 μ/div

Figure 4.10 PWM-related DC bus voltage ripples during nonconducting state of the rectifier—left side (*A*) and during conducting state—right side (*B*).

During the state of the FC when the input rectifier is temporarily unloaded, its DC output current is zero (intervals of type *A*), the DC bus voltage unbalance in relation to ground changes relatively rapidly only when the state of a transistor of the output inverter is changed. Within time intervals between transistors switching, DC bus voltage unbalance remains almost unchanged, which consequently results in practically rectangular shape of DC bus CM voltages $V_{(DC+)}$ and $V_{(DC-)}$ (Figure 4.10—left side). Nearly rectangular changes of DC bus CM voltages are tightly synchronized with the output-side CM currents $I_{CM,OUT}$, whereas at the same time the input CM currents $I_{CM,IN}$ are not observed as they are blocked by all nonconducting diodes of the rectifier bridge.

At the right-hand side waveforms presented in Figure 4.10, interval *B*, DC bus CM voltages are changing continuously in relatively slow speed in relation to PWM originated voltage ripples. These slow changes in DC bus CM voltages are a result of loading process of DC bus capacitance by conducting diodes in a six-pulse rectifier bridge. During this time interval where DC bus is connected via conducting diodes with power grid, we can observe significant CM currents flowing at both sides of FC, input $I_{CM,IN}$ and output $I_{CM,OUT}$.

During time interval *B*, when two of the diodes in the rectifier are in conducting state, after each DC bus CM voltage change caused by output voltage switching, unequally charged parasitic capacitances of DC bus $C_{DC+,Gnd}$ and $C_{DC-,Gnd}$ are reloaded in order to equalize V_{DC} voltage distribution between them. Waveform of CM input current during rectifier conduction state is presented in Figure 4.10 (right side). The shape of this current is closely correlated

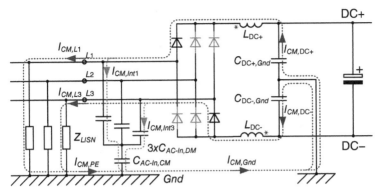

Figure 4.11 Grid-side CM currents propagation paths in loaded phases.

with PWM pattern; nevertheless, the current ringings observed after each com-mutation process are much slower in relation to the adequate ringing of the CM current at the output side. This is an effect of longer time constant of reload-ing circuit. Simplified circuit model of adequate current paths is presented in Figure 4.11.

Each commutation process in the inverter output changes slightly CM DC bus voltages unbalance in relation to ground, which was visible as rectangular shape of CM voltages while the input rectifier diodes were not conducting. If two of the rectifier's diodes are in conducting state due to the DC bus capacitance loading process, the DC bus CM voltage unbalance is possible to be equalized via the input side of FC. To equalize slightly different CM voltages at capacitances $C_{DC+,Gnd}$ and $C_{DC-,Gnd}$ with DC bus voltage V_{DC} (Figure 4.11) (4.3), one of the capacitances should be more charged and the other partly discharged.

This process generates two currents, $I_{CM,L1}$ and $I_{CM,L3}$, flowing via two currently conducting diodes and the power grid impedances toward ground (Figure 4.11). CM impedances of the power grid in HF frequency range sub-stantially depends on line-to-ground capacitances that can significantly reduce their values. Line-to-ground capacitances of the power grid are usually non-constant and difficult to predict; therefore, in order to get rid of that effect, the line impedance stabilization network (LISN) are often used. In the proposed test setup, the grid impedance is represented by Z_{LISN}, which equalizes and stabilizes grid impedance changes.

In practical applications at the input side of FC, CM capacitors are often added for decreasing CM impedance seen from the point of view of the rectifier AC side. These capacitances cause part of a CM current generated by FC to be shunted to ground inside FC and therefore the external emission of ASD can be decreased. In the presented model in Figure 4.11, the total CM current

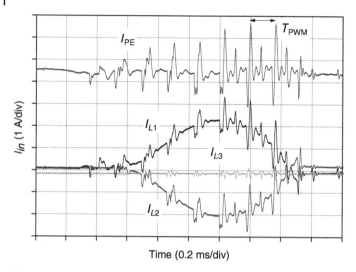

Figure 4.12 Input-side CM current waveform during single pulse of line current of the rectifier operating in noncontinuous current mode.

generated by FC in phase $L1$ consists of internal part $I_{CM,Int1}$ and external part $I_{CM,L1}$ and respectively for other phases.

CM current components recorded in all power lines $I_{CM,L1}$, $I_{CM,L2}$, $I_{CM,L3}$ and the grounding connection between the power grid and FC $I_{CM,PE}$ allow analyzing their distribution during commutation processes. According to the presented experimentally recorded waveforms (Figure 4.12), initially, just after the commutation process, the input-side CM currents flow primarily in two phases with nonzero value of LF current component of power line frequency, in the same directions. Therefore, the sum of this two CM currents $I_{CM,L1}$ and $I_{CM,L2}$ is returning back to the FC via grounding connection between the power grid and FC realized using PE grid terminal (4.6).

$$I_{CM,L1} + I_{CM,L2} = I_{CM,PE} \qquad (4.6)$$

CM current observed in third, currently nonconducting, phase is significantly lower. Differences in values of CM components of currents in conducting and nonconducting phases can be explained based on the circuit diagram presented in Figure 4.13. Similarly as for conducting phases (Figure 4.12), charging and discharging currents $I_{CM,DC+}$ and $I_{CM,DC-}$ are flowing toward the power grid, but to enter to the nonconducting phase (in the analyzed case $L2$) they have to overcome currently nonconducting diodes. Parasitic capacitances of diodes in blocking state are much lower than in conducting state; therefore, the impedance for high-frequency CM currents is relatively much higher. The

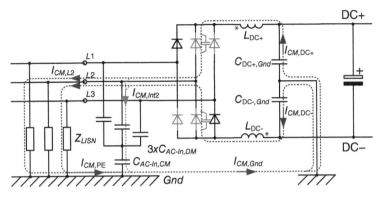

Figure 4.13 Grid-side CM currents propagation paths in nonconducting phase.

recorded CM current in nonconducting phase can also be increased in relation to the value resulting from diodes' capacitances, because many of other small parasitic capacitances, for example, capacitance of connecting wires, have comparable values and can become significant.

4.7 Summary

Internal parasitic capacitances of components of an FC affect mainly propagation of conducted emission generated at the motor side toward the power grid. Some of these capacitances can be assigned to equivalent lumped capacitances associated with the input and output terminals of FC; thus, they can be analyzed together with other external parasitic and intentional capacitances existing at both the input and output sides of the FC.

The most significant for conducted emission propagation, internal parasitic capacitances that cannot be assigned to the FC external terminal can be associated with the positive and negative DC bus connections. DC bus-to-ground capacitances, parasitic as well as with purpose, determine mostly the efficiency of propagation process of PWM originated CM voltages and current transients appearing in the DC bus connection. Levels of those ripples and inductances of DC bus connections essentially affect the transfer ratio of conducted emission from the motor side to the grid side of FC.

Therefore, conducted emission filtering components introduced into DC bus, CM chokes, and balancing capacitors can have significant effect on CM voltages and currents propagation inside FC and should be extensively considered at the design stage of external line filters. Conducted emission filtering components used at DC bus level have to be cautiously optimized because oversizing of these components can also lead to opposite effects, for example, increased emission due to resonances possible to appear. Likewise, CM capacitances implemented into DC bus besides positive effects for conducted emission propagation will also increase FC's output and input CM current.

The fundamental problem with effective optimization of conducted emission filtering at the DC bus level is associated with significant influence of parameters of a grid connection and FC's load within high and wide frequency range. Therefore, it is a severe difficulty to design adequate and efficient DC bus filters for applications where HF impedance characteristics of load and power grid vary significantly.

References

1 R. Smolenski, "EMI measuring procedures in smart grids," in *Conducted Electromagnetic Interference (EMI) in Smart Grids*, Springer, pp. 145–147, 2012. Available at http://link.springer.com/chapter/10.1007/978-1-4471-2960-8_6

2 L. Xing and J. Sun, "Motor drive common-mode EMI reduction by passive noise cancellation," in *Proceedings of the 2011 14th European Conference on Power Electronics and Applications*, Aug. 2011, pp. 1–9.

3 F. Zare, *Electromagnetic Interference Issues in Power Electronics and Power Systems*, F. Zare, Ed., Bentham Science Publishers Ltd., 2011. ISBN: 978-1-60805-240-0. Available at http://www. benthamscience.com/ebooks/contents.php?JCode=9781608052400

4 S. Mandrek and P. J. Chrzan, "Quasi-resonant dc-link inverter with a reduced number of active elements," *IEEE Transactions on Industrial Electronics*, vol. 54, no. 4, pp. 2088–2094, Aug. 2007.

5 T. Qi and J. Sun, "DC bus grounding capacitance optimization for common-mode EMI minimization," in *2011 Twenty-Sixth Annual IEEE Applied Power Electronics Conference and Exposition (APEC)*, March 2011, pp. 661–666.

6 D. Zhao, J. Ferreira, A. Roc'h, and F. Leferink, "Common-mode DC-bus filter design for variable-speed drive system via transfer ratio measurements," *IEEE Transactions on Power Electronics*, vol. 24, no. 2, pp. 518–524, 2009.

7 X. Fang, S. Li, and D. Jiandong, "Prediction model of conducted common-mode EMI in PWM motor drive system," in *2010 First International Conference on Pervasive Computing Signal Processing and Applications (PCSPA)*, Sept. 2010, pp. 1298–1301.

8 M. Jin and M. Weiming, "Power converter EMI analysis including IGBT nonlinear switching transient model," in *Proceedings of the IEEE International Symposium on Industrial Electronics, 2005 (ISIE'05)*, vol. 2, June 2005, pp. 499–504.

9 T. Liu, Y. Feng, R. Ning, T. T. Y. Wong, and Z. J. Shen, "Extracting parasitic inductances of IGBT power modules with two-port s-parameter measurement," in *2017 IEEE Transportation Electrification Conference and Expo (ITEC)*, June 2017, pp. 281–287.

10 A. Domurat-Linde and E. Hoene, "Analysis and reduction of radiated EMI of power modules," in *2012 7th International Conference on Integrated Power Electronics Systems (CIPS)*, 2012, pp. 1–6.

11 T. Qi, J. Graham, and J. Sun, "Characterization of IGBT modules for system EMI simulation," in *2010 Twenty-Fifth Annual IEEE Applied Power Electronics Conference and Exposition (APEC)*, , IEEE, 2010, pp. 2220–2225. Online.. Available at http://ieeexplore.ieee.org/xpls/abs_all.jsp?arnumber=5433545

12 A. Videt, P. Le Moigne, N. Idir, P. Baudesson, and X. Cimetiere, "A new carrier-based PWM providing common-mode-current reduction and dc-bus balancing for three-level inverters," *IEEE Transactions on Industrial Electronics*," vol. 54, no. 6, pp. 3001–3011, 2007.

13 D. Zhao, J. A. Ferreira, H. Polinder, A. Roc'h, and F. B. J. Leferink, "Noise propagation path identification of variable speed drive in time domain via common mode test mode," in *2007 European Conference on Power Electronics and Applications*, 2007, pp. 1–8. Available at http://ieeexplore.ieee.org/stamp/stamp.jsp?arnumber=4417580

14 D. W. P. Thomas, C. Christopoulos, F. Leferink, and H. Bergsma, "Practical measure of cable coupling," in *International Conference on Electromagnetics in Advanced Applications, 2009 (ICEAA '09)*, pp. 803–806, 2009.Available at http://ieeexplore.ieee.org/stamp/stamp.jsp?arnumber=5297315

15 L. Xing and J. Sun, "Conducted common-mode EMI reduction by impedance balancing," *IEEE Transactions on Power Electronics*, vol. 27, no. 3, pp. 1084–1089, March 2012.

16 S. Czapp, "The impact of higher-order harmonics on tripping of residual current devices," in *13th Power Electronics and Motion Control Conference, 2008, EPE-PEMC 2008*, Sept. 2008, pp. 2059–2065.

17 S. Czapp and J. Guzinski, "The effect of the motor filters on earth fault current waveform in circuits with variable speed drives," in *International School on Nonsinusoidal Currents and Compensation, 2013 (ISNCC'13)*, June 2013, pp. 1–6.

18 G. L. Skibinski, R. M. Tallam, M. Pande, R. J. Kerkman, and D. W. Schlegel, "System design of adjustable speed drives, part 1: equipment and load interactions," *IEEE Industry Applications Magazine*, vol. 18, no. 4, pp. 47–60, July 2012.

5

Modeling of Conducted Emission in ASD

> *Essentially, all models are wrong, but some are useful.*
> George E.P. Box

The primary objective of computational modeling of applications with power electronics converters is to obtain as many functionally essential characteristics of evaluated construction as possible without the necessity to build a prototype. Designing processes of power electronics converters are usually preceded by in-depth simulation analysis of fundamental behavior that allows predicting most of the significant issues related to fundamental behavior before building physical model, but very rarely allows solving problems related to EMC performance entirely effectively.

In commonly used design process of ASD applications, conducted emission issues are usually addressed after systems installation. Adding extra components for limiting conducted emission at this stage very often significantly affects costs and completion time, or even requires redesigning of entire system. Therefore, methodologies that allow predicting the EMC problems as early as possible are developed very extensively [1–3].

Effectiveness of currently available simulation tools used for power electronics converters design is systematically increasing and nowadays it is quite satisfactory for assessing fundamental functionalities of converters related to the used modulation methodologies. Nevertheless, detailed analysis of transistors and diodes commutation processes usually require consideration of parameters of parasitic couplings existing internally in power electronic components and also resulting from externally connected subcomponents of ASDs that significantly limit reliability of obtained results [4,5].

Commonly, even in regular analysis of power electronic converters, extension of frequency bandwidth above few hundreds of kilohertz results in significant

High Frequency Conducted Emission in AC Motor Drives Fed by Frequency Converters: Sources and Propagation Paths, First Edition. Jaroslaw Luszcz.
© 2018 by The Institute of Electrical and Electronic Engineers, Inc. Published 2018 by John Wiley & Sons, Inc.

increase of influence of parasitic couplings, which cannot be omitted. Analysis of EMC behavior of FC extends model complexity excessively because it requires taking capacitive couplings to ground into account that are essential for conducted emission phenomena analysis. Parasitic couplings between energized and grounded components of converters are omnipresent, numerous, and distributed in all subcomponents, including nonelectrical conducting construction [6].

Complexity of analysis of conducted emission in ASDs is even higher, because ASDs in addition to FCs include external and usually significantly changeable loads. The motor as a converter load is a relatively bulky component with long winding wires and grounded stator placed closely to each other, thus resulting in strong capacitive couplings to ground. The motor feeding cable is often of a considerable length in comparison to the converter and motor dimensions and can be variantly placed in relation to other cabling—especially grounded cable trays. Therefore, a kind of the motor cable, cable shielding exists and cable arrangement in particular application can influence significantly the overall EMC performance of the whole ASD application [7].

Modeling of EMC behavior of an ASD as the whole system, consisting of an FC, an AC motor, the motor feeding cable, and optionally EMI filtering components, requires knowledge of broadband characteristics of all subcomponents. Most of technical specifications of commonly used ASD subcomponents currently provided by manufacturers primarily do not include wide-band specifications that are detailed enough for simulation analysis, even necessary within the conducted emission frequency range. Therefore, modeling of EMC requires determining missing parameters that describe broadband behavior of all subcomponents. Identification of all those missing parameters can be carried out in two ways: by measurement using external terminals only or by intrusion into the inside of components. Both the methods are problematic. Terminal-based approach is much easier to run experimentally but allows acquiring only strongly limited information about internal structures of identified components. On the other hand, access to internal parts of identified components is much more elaborate and hazardous for tested devices, and thus not always possible; however, it can be very significant for effectiveness of model simplification [8,9].

Summarizing, EMC analysis of ASDs is much more extensive, complex, and challenging compared to the FC as one of the ASD components. Detailed investigation of the influence of all parasitic couplings existing in ASD leads to enormous expansion of model complexity, thus consequently greatly increasing computational overhead and efforts necessary for identification process of numerous parameters of such models. The use of very extensive models usually also rises uncertainty about the accuracy of obtained results; therefore, reasonable optimization and simplification of models is one the most expected outcomes of current developments in EMC analysis. In spite of many difficulties, modeling simulation study of conducted emission in ASDs is increasingly intensively developed over the last years. In a number of published papers, new methodologies for precise and successful solving of particular EMC issues occurring in contemporary applications of ASDs have been reported [10–12]. Nevertheless, maintaining reasonable balance between model complexity and its effectiveness is still the greatest challenge in current research areas related to EMC.

5.1 Parasitic Couplings as the Most Meaningful Factor of Conducted Emission Generation

Parasitic couplings existing in all electrical circuits are related to magnetic and electric components of fields generated by alternating currents and voltages and cannot be eliminated entirely. Therefore, all electrical circuits can be influenced, and possibly interfered, by signals generated by external electrical and magnetic field components originated from other nearby circuitries and transferred by means of parasitic couplings. Effectiveness of both types of parasitic couplings increase with the increase of frequency of considered signals. From the general point of view of EMC, both parasitic couplings—by magnetic and electric field components—can be likewise meaningful [13,14]. In ASDs, the initial source of HF emission is fast switching of voltages and consequently also HF current transients generated at the output terminals of FC. The rate of change of AC motor operational currents (Figure 5.1) is strongly limited by relatively high inductances of motor windings; therefore, spectrum contents of generated magnetic field associated with motor current is also limited to frequencies only slightly higher than the PWM carrier frequency. Relatively low steepness of operational motor current (dI/dt) results in magnetic field-related emission that can be significant only in the low part of frequency range of conducted emission.

Furthermore, LF components of output currents of FC correlated with the motor operating current are predominantly DM, which means that related current path for these components are well defined by power wires and low DM impedances of motor windings. Areas of loops formed by DM output currents are usually of much smaller area than for CM currents, because of the use of three wires cabling, where wires of each phase are very tightly arranged with wires of other phases. Apart from the unique cases, contribution of motor operational current ripples to overall intensities of interference originated by means of magnetic field couplings is usually not significant, because of relatively low frequency of the dominating spectral content. Therefore, this type of magnetic field emission is usually not emphasized in typical ASD applications.

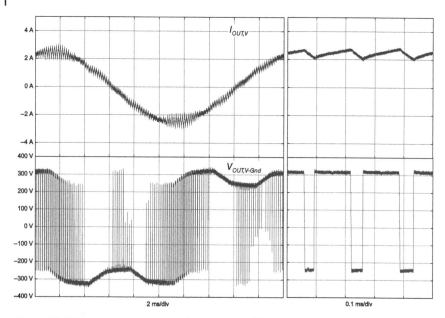

Figure 5.1 Voltage and current waveforms at one of the converter's output terminals.

Capacitive parasitic couplings occurring in ASDs result in more significant effects in relation to magnetic fields generated by drive operating currents. First, output voltage changes at FC terminals are significantly higher in magnitudes, more than 100% of nominal voltage, in relation to the ripples of the motor's operational currents that usually do not exceed 10Second, output voltage changes are much faster (Figure 5.1), because they are limited only by small parasitic capacitances. Thus, voltage steepness easily follow transistor switching characteristics, changes of which are often much faster than 1 μs.

Rapid changes of output voltages as a primary reason, together with omnipresent parasitic capacitances as necessary second condition, are the dominant cause of HF parasitic leakage currents. These parasitic transient leakage currents add themselves at the FC output to motor operating currents with much slower ripples, and then return back to the converter via parasitic capacitances, spreading out into many possible return paths. Simplified example of parasitic HF capacitive current transients distribution at the motor side of ASD is presented in Figure 5.2.

According to the presented parasitic output current distribution at the motor side of ASD, total phase current of FC output $I_{OUT,V}$ is a sum of the motor operating current $I_{OUT,V1-V2}$ and parasitic currents: one CM—$I_{CM,V-Gnd}$ and two DM—$I_{DM,V-U}$, $I_{DM,V-W}$ (5.2).

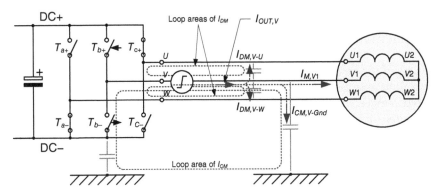

Figure 5.2 Converter's current subcomponents at one of the output terminals.

$$I_{OUT,V} = I_{OUT,V1-V2} + I_{CM,V-Gnd} + I_{DM,V-U} + I_{DM,V-W} \tag{5.1}$$

Both, CM and DM parasitic capacitive couplings together with CM and DM output voltage rapid changes have an effect on conducted emission generation as HF current transients. Nevertheless, CM current components are usually the most meaningful for two reasons. First, capacitive couplings between energized and grounded components are usually stronger than between components itself, because of greater amount and size of grounded parts. Second, current loop areas formed by HF CM current transients, which are essential for conducted emission generation, are usually significantly greater than for DM currents (Figure 5.2). CM currents flow via not exactly defined paths created by different grounded components and associated CM parasitic capacitances, whereas DM currents have well-defined paths by relatively low interphase impedances. Third, much higher phase-to-ground impedances than phase-to-phase impedances cause that CM currents are not damped as efficiently as DM currents.

Summarizing, from the point of view of conducted emission generation in ASD, values of parasitic capacitances, their allocation across ASD components, as well as distribution into CM and DM categories are the most essential. Therefore, the most challenging task in modeling conducted emission is successful determination of parasitic capacitances and its distribution. Thus, the extension of the complexity of applied models should be closely correlated with technical feasibility to identify their parameters with the adequate accuracy and with the reasonable overheads.

5.2 Difficulties in Determination of Parasitic Capacitive Couplings

Many of the methods widely used for analysis of capacitive couplings in electrical circuits are based on representation of these couplings by equivalent capacitances. Circuit representation of parasitic couplings allow using circuit-oriented simulation tools, such as Spice, that can significantly simplify further analysis. Integrating advantages of full-wave electromagnetic methods for problem formulation and circuit simulation methods for extensive analysis allow to solving efficiently complicated field geometries with smooth transition to lumped circuit element representation [15]. However, detailed formulation of internal dependencies using FEM methods requires elaborate specification of inner geometries of subcomponents of the analyzed device [16–18]. Therefore, such approach can be most efficient at device design stage, and thus less successful at the application stage in which access to internal subcomponents is strongly limited. In analysis of ASD applications, where many of the components already have a completed design process and fixed construction, this method can be effectively used only when sufficiently detailed models along with adequate parametrization procedures are provided by manufacturer.

Frequency range of a typical analysis of conducted emission conventionally should cover a frequency range of at least 9 kHz to 30 MHz. Such a frequency range is wide enough for revealing more complex behavior of even the simplest passive components like resistor, inductor, and capacitor, which expose lossy, capacitive, and inductive, character simultaneously. Currently, not many of the manufacturers provide parametrized broadband models for components used in the ASDs, especially of high power, for example, AC motors, inductors, and FCs. Even power cables recommended for ASD applications are usually not comprehensively specified by manufacturres from the point view of frequency range of conducted emission analysis, despite the fact that cable specification methodology is already well defined and commonly used in signal transmission applications.

Due to many insufficiencies in broadband models availability of components used in ASDs, applications developers can rely only on manufacturer's general EMC-related recommendations that are provided usually only for the most common cases. Otherwise, if such recommendations are not sufficient, wide-band identification methods need to be used for better identification of components' characteristics in a wide frequency range. First, the identification methods that do not require too much intrusion into the inside structure of the component are usually preferred. Identification methods are available that allow successful characterization of electrical devices in a wide frequency range based on terminals referenced measurement of impedance–frequency characteristics, for example, N-port network approach [19,20].

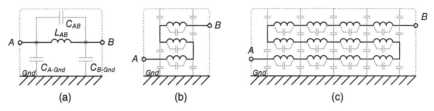

(a) (b) (c)

Figure 5.3 Different levels of complexity of circuit models used for analysis of iron core inductors. (a) With equivalent circuit-lumped components connected to inductor's terminals only. (b) With distributed equivalent parasitic capacitances correlated with winding's layers. (c) With parasitic capacitances distributed to each winding's turn.

The main disadvantage of a device ports approach, from the point of view of analysis of EMC issues, is that in models obtained using this method, significant part of knowledge about internal behavior and physical structure of the identified device cannot be represented adequately. As a consequence, in many cases only "black box" representation is possible to obtain, for which the possibility of formulation of equivalent circuit model of its internal structure is significantly limited. Therefore, circuit models possible to obtain with the use of this method are usually limited to equivalent circuit components directly connected to model's terminals. In analysis of conducted emission, distribution of internal parasitic capacitances inside components is meaningful; therefore, terminal circuit models are substantially less effective.

An example of gradation of complexity level of circuit models used for representation of distributed parasitic capacitances can be conceivably explained based on a simple winding component that can be an iron core inductor (Figure 5.3). Sketched overview of commonly applied constructions of a wounded component imply the existence of parasitic capacitances between each nearby turns of windings and between boundary turns and the ground.

Modeling of parasitic capacitances distributed between all winding's turns requires the use of field-oriented methods to determine partial components of capacitances correlated with partial inductances of each single turn of the winding. Examples of adequate configurations of circuit models are presented in Figure 5.3c. On the other hand, the same inductor can be modeled in much more simplified form, by replacing all partial parasitic capacitances with one equivalent parasitic self-capacitance C_{AB} referenced to the inductor terminals A and B and two parasitic capacitances representing couplings between the winding and the ground connected to each inductor terminals and the ground (Figure 5.3a).

In such a considerably simplified approach, all equivalent capacitances are connected only to inductor's terminals and therefore can be easily determined by measuring impedances via externally accessible terminals. Depending on the particular internal structure of the modeled inductor and frequency range

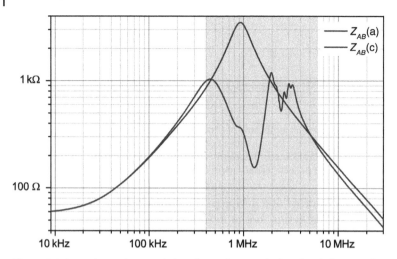

Figure 5.4 Impedance characteristics of an inductor calculated with the use of circuit models of different complexity: Z_{AB} model a—parasitic capacitances represented by 3 equivalent capacitances (Figure 5.3a) and Z_{AB} model c—by 30 equivalent capacitances (Figure 5.3 c).

of performed analysis and required accuracy, one of the presented approaches can be chosen for adequate representation. Therefore, in a number of cases excessive extension of number of model's details can be unreasonable, thus a level of detail of the model should be carefully considered and justified for each particular case, based on the preliminary verification of complexity of impedance characteristics. For modeling of winding components, compromising solution can also be used quite often, where determination of only selected essential parasitic equivalent capacitances of modeled windings is sufficient. Rough overview of mechanical configuration of inductor's windings focusing on winding's sectioning and layering can be helpful for estimation of the complexity of parasitic capacitances distribution. The initial estimation of the expected complexity level of parasitic capacitances distribution in wound component can also be done based on rough overview of selected impedance characteristic that usually well reflects internal behavior of components. Such initial diagnosis can be helpful for choosing the adequate complexity level of circuit model configuration and allows decreasing identification efforts.

Comparison of impedance–frequency characteristics of exemplary winding of an inductor consisting of three layers are presented in Figure 5.4. Impedance characteristics that are shown have been calculated for circuit model consisting of only three equivalent capacitances (Figure 5.3a) and for the model with 30 capacitances arranged according to Figure 5.3c.

Presented characteristics show that the entire range of frequencies where successful application of discussed models is possible can be divided into three subranges. First subrange is the lower part of conducted emission frequency range where inductive character of component is dominating and thus impedance increases logarithmically with the frequency. In this frequency subrange, whose upper boundary can vary depending on properties of particular component, taking into account parasitic capacitances is usually not essential.

Second subrange is the middle part of conducted emission frequency range where parasitic capacitive couplings existing in windings are manifested by multiresonance effects. Depending on the intensity of resonance phenomena, precise modeling of the inductor within this range is much more critical and requires a detailed specification of distribution of internal parasitic couplings.

Third subrange is the upper part of conducted emission frequency range, where, according to the presented exemplary simulation results, capacitive behavior of inductor is dominating. In this frequency subrange, component's impedance should decrease logarithmically with the increase of frequency, as shown in Figure 5.4 for frequencies above about 4 MHz. Nevertheless, in this frequency range, impedance characteristics obtained experimentally usually vary significantly and irregularly, which makes the choice of adequate circuit model configuration and identification of its parameters extremely difficult, especially for components with low internal losses and parasitic magnetic couplings. In this frequency range, circuit model approach cannot be sufficiently effective and other methods such as vector fitting can be recommended [21,22].

An example of influence of the level of circuit model complexity on the estimated insertion loss frequency characteristics of the evaluated inductor within conducted emission frequency range is presented in Figure 5.5. Presented insertion loss characteristics have been calculated for the same inductor modeled by different circuit models presented in Figure 5.3a and c with adequate impedance characteristics presented in Figure 5.4.

Inadequacy exposed on impedance characteristics in the middle of the frequency range (Figure 5.4) is also reflected in insertion loss characteristics calculated based on the proposed circuit models for the modeled inductor as an EMI filter presented in Figure 5.5. Calculated attenuation of inductor as an EMI filter is mostly dependent on the model accuracy mainly within the middle of the frequency range, where internal resonances caused by distributed parasitic capacitances occur.

Summarizing, difficulty in adequate representation of component's broadband behavior by a circuit model and efforts accompanied with its parameter identification strongly increase with the increase of frequency range of analysis. Unfortunately, frequency range up to 30 MHz, commonly used for conducted emission analysis, is wide enough for parasitic resonance effects to occur in majority of high-power components used in ASDs. Nevertheless, depending on the complexity of broadband behavior of particular components under analysis, circuit modeling can be used effectively and its results can be very relevant in investigation, because it allows obtaining rational balance between model

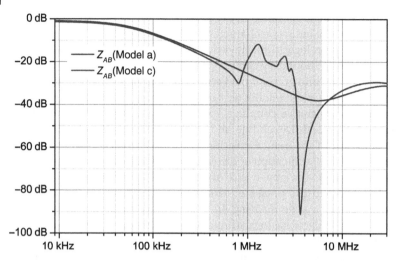

Figure 5.5 Insertion loss of the inductor calculated based on circuit models of different complexity: Z_{AB} model a—parasitic capacitances represented by 3 equivalent capacitances (Figure 5.3a) and Z_{AB} model c—by 30 equivalent capacitances (Figure 5.3c).

complexity and achieved accuracy. Furthermore, even highly simplified circuit models can be more profitable than more accurate "black box" because they allow better understanding of the analyzed phenomena, indicating the most meaningful ones, and also allow more efficient use of such models for further improvements.

5.3 Considerations on Simplified Simulation of Conducted Emission Generation in ASD

5.3.1 Limitations of Bandwidth in Circuit Modeling

In analysis of EMC, commonly considered spectrum of conducted emission covers the frequency range up to 30 MHz. Modeling of power electronic applications in such a wide frequency range results in necessity to use broadband models that are capable to reflect a device behavior associated with parasitic couplings. In majority of high-power devices used in power electronic applications, parasitic resonances appear below 30 MHz; it thus can have significant influence on generated conducted emission. Therefore, broadband approach is usually highly required, although it is much more complicated.

Circuit models widely used for simulation of power electronic applications are very effective, but mostly within limited upper frequency range, which is tightly correlated with minimum frequency of parasitic resonance inceptions.

Frequency range of usefulness of simple circuit models used in LF analysis can be relatively easily extended—slightly above frequency of the first resonance—especially successfully for devices where the lowest-frequency resonance is expressly dominating.

Further extension of frequency range where adequacy of terminal referenced circuit models can be acceptable require to use much more complex and laborious identification methods. In many cases, precise enough identification of parameters of wideband circuit models becomes extremely difficult, when intrusion into the interior of device is not accessible. Significant difficulties in identification of parameters of wideband circuit models encourage or even pressurize for using reasonable simplification.

It is possible to build mathematical models reliable in a wide frequency range based on impedance–frequency characteristics measured externally, seen via device terminals only. Such models are known as "black box" and "N-port" models and can precisely reconstruct device behavior in a wide frequency range; nevertheless, they do not grant smooth possibility for transformation into beneficial circuit representation. Transition of those models into circuit representation is possible using mathematical methods; nevertheless, results obtained are usually not enough correlated with modeled physical phenomena. Lack of analytical benefits resulting from physical interpretation of obtained circuit models significantly reduces their applicability.

Summarizing, modeling of ASDs within a wide frequency range of DC using circuit models is still very challenging and so far obtainable only for limited scope of the applications and limited upper frequency range. Striving to complete modeling of a whole ASD with the use of single broadband model is very demanding, laborious, and for most cases unreasonable, because of too wide range of essentially different phenomena that have to be precisely represented to obtain beneficial results. The most beneficial results of simulation are achieved when the developed model not only precisely emulates behavior of the modeled application in a particular mode of operation but also allows running simulation experiments by changing crucial parameters and obtaining still reliable results.

5.3.2 Characteristic Frequency Subbands Present in ASDs

Few characteristic frequency subbands related to individual phenomena can be distinguished in the analysis of ASDs in frequency domain, which in a different way influence ultimately generated conducted emission (Figure 5.6). These are as follows:

- Grid frequency and FC output frequency subrange with accompanied higher order harmonics—up to few kilohertz.

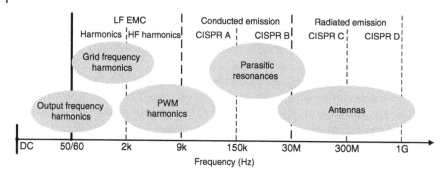

Figure 5.6 Characteristic subranges of frequencies essential for generation of conducted emission in ASDs.

- PWM frequency subrange with accompanied higher order harmonics— between few kilohertz and several of tens of kilohertz.
- Parasitic resonances frequency subrange—above few hundreds of kilohertz.

From the point of view of FC operation, frequency subband related to PWM carrier frequency is fundamental, which in ASD applications is usually located within the range of few to few tens of kilohertz, depending on the rated power. Higher order harmonic components initially generated by PWM frequency can achieve meaningful amplitudes within frequency range up to few tens of kilohertz. The grid frequency and the FC output frequency, which are much lower than low limit of considered conducted emission range, can also have indirect influence on the final characteristics of conducted emission.

From the point of view of generation of conducted emission, subbands related to parasitic resonances, are tightly correlated to parasitic capacitances energized by rapid changes of commutated voltages that initiate CM currents.

Methods of analysis of fundamental operation of FC within PWM frequency range are already well developed and successfully used [23]. Nevertheless, the most effective methods based on equivalent circuit models are difficult to expand toward higher frequencies, because of difficulties in precise identification of parameters of parasitic couplings in HF range.

5.3.3 Variation of Factors Influencing Conducted Emission in ASD

Conducted emission of a permanently installed ASD is not steady; it is sensitive to many parameters that usually change their values during typical operation. In ASDs, PWM as the most dynamic and influencing process is dependent on a number of parameters such as controlled speed, drive load changes, power supply fluctuations, and autotuning procedures of modulation pattern. Therefore, this can result in significant variation of PWM-originated conducted emission.

Transistors switching dynamic characteristics are also dependent on parameters of the FC load; therefore, parameters of output voltage switchings usually continuously fluctuate because of cyclic changes of magnitudes of motor operating currents according to controlled output frequency.

Conducted emission can also be permanently modified at the installation stage of ASD due to introduced, intentionally or not, differences in arrangement of drive components and the quality of interconnections that can change wide-band characteristics of real and parasitic impedances. The length of the motor feeding cable, which is most often exclusively dependent on particular application requirements, is also the foremost crucial parameter. Taking into consideration all existing influences, the complexity of analysis expands enormously; therefore, in every analysis, reasonable selection of most of the significant factors is usually necessary for successful results.

5.3.4 Generation and Propagation of Conducted Emission in ASD

Despite the fact that conducted emission finally generated by ASD is strongly dependent on propagation processes, it is very helpful for better understanding of conducted emission phenomena, to separate all the meaningful processes into two groups: one group of phenomena related to generation processes only, and the other group related to propagation processes only.

Generation of conducted emission is initiated by rapid voltage changes at converter's outputs, with high rate of steepness, repeated quasi-regularly according to the PWM control pattern (Figure 5.6). Voltage spectral density of such a source can be represented as equivalent voltage source $V(f)$, with spectral content determined using Fourier transform of voltage waveforms at the output terminals of FC, or estimated according to the simplified methods presented in Chapter 2. Rapid voltage changes generated at the output terminal of FC are transmitted in the limited speed toward motor terminals via single wire of the motor feeding cable and further along motor's windings wires, reloading all encountered parasitic capacitances from previous voltage potential to the opposite, for example, from V_{DC-} up the V_{DC+}. Due to the distributed character of parasitic capacitances and remarkable time delays in the relatively long output circuit, CM impedance–frequency characteristic of FC load is not purely capacitive and usually exposes a number of internal resonances.

CM currents generated this way are the primary origin of conducted emission of ASDs, although highly significant part of the generation processes are also phenomena related to propagation processes occurring at the motor side of FCs, which can affect spectral content of the resultant output CM current $I_{CM,Out}$. Therefore, generation process of conducted emission, from the point of view of a whole application with an ASD, is initiated at the output side of FC but is also dependent on many other phenomena, including propagation processes of CM currents at the output side of FC.

CM currents generated at the motor side of ASD are a meaningful indicator of general EMC performance, despite the fact that its limitation of conducted emission at the output side of FC is not directly covered by recommendations of EMC standards. Output-side CM currents affect electromagnetic environment by near-field emission, which is a common source of interference generated by crosstalk between nearby cable paths.

The most important consequence of CM currents generated at the motor side of FC is their propagation toward the grid side of FC through DC bus connections and injection as a conducted emission into the power grid. All phenomena related to this process, from the point of view of an ASD, can be classified as propagation, because CM currents spectrum already formed at the output side is only transmitted into grid side of FC due to existing propagation paths, both intentionally created and parasitic.

The proposed classification of HF processes occurring in ASDs into generation and propagation categories allows underlining the importance of EMC performance at the motor side of ASD, which is not often adequately highlighted in developments of limitations of conducted emission at the grid side of FC. Quality of HF performance of the motor side of FC is mostly analyzed from the point of view of limitation of CM current flows in the AC motor. Independent analysis of CM currents generation process can also be helpful for simplification of wide-band identification methods of component's parameters that affect the generation process such as the length of motor cable.

5.3.5 Foremost Objectives of Conducted Emission Simulation Analysis

The primary objective of analysis of conducted emission in ASDs is estimation of spectral characteristic of voltages generated at the terminals of FC connected to the power grid, which are an essential indicator of potential interfering effectiveness of other devices connected to the same grid. According to the established standards related to conducted emission measurement, peak magnitudes of CM transient currents injected into the power grid and their repetition frequency are the main determinant of conducted emission levels.

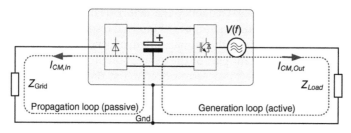

Figure 5.7 General schematic of conducted emission generation and propagation in ASD.

The above specification indicates that analysis of all possible interactions in detail in time domain is not necessarily needed, because the most meaningful phenomena predominantly decide about the final result. This constraint makes it possible to introduce really great simplification, which is alteration from focusing on all possible factors that really affect, but move toward finding only the most essential factors that allow successful estimation of conducted emission in the worst-case scenario.

Second, but not less important, objective of analysis of conducted emission in ASDs is to deliver for developer of ASD applications an efficient tool that makes possible to estimate the influence of varying components of ASD on resultant conducted emission. A motor cable is one of ASD components that is highly individualized for each application, because of its varying length. From the point of view of conducted emission limitation in ASDs, the maximum values of emission levels as well as frequency ranges of their occurrence are the most important.

5.3.6 Feasible Advantageous Simplifications

As the analysis of all HF processes occurring in ASDs within the conducted emission frequency range up to 30 MHz is very difficult, extremely laborious, and in many cases not reasonable, a range of simplified approaches is usually implemented. Based on the considerations discussed in previous section, two concepts of possible powerful simplification have been indicated.

First, CM currents can be analyzed separately, without simultaneously taking into consideration DM operational currents in ASDs. This simplification can be successfully used especially for analysis of generation processes of CM currents at the motor side of FC, and is presented in Section 5.4.

Second, CM currents are generated as repeated transients accompanying reloading processes of relatively small parasitic capacitances, which are usually shorter in time than modulation period. Detailed analysis of system response to a single-voltage transient and taking into consideration repetition frequency of these transients allow obtaining acceptable approximation of expected conducted emission with a considerable reduction of simulation overheads. This approach is described in Section 5.5.

5.4 Single-Loop Representation Method

First, each voltage switching at the output terminals of FC is a reconnection of converter's terminals from one of DC bus potentials to the opposite one in much shorter time than AC motor windings time constant and modulation period. Therefore, operating currents in motor windings cannot change significantly during relatively short commutation process. Examples of current waveforms

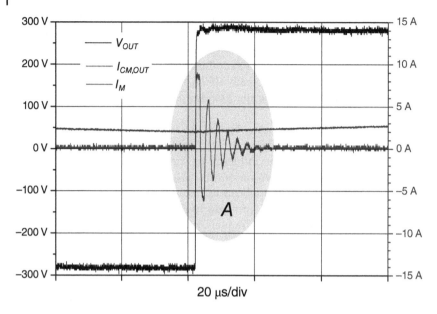

Figure 5.8 Output voltage V_{OUT}, CM current $I_{CM,OUT}$, and motor current I_M waveform during typical commutation process at converters output terminal.

during single switching process are presented in Figure 5.8 and in a wider view in Figure 5.1.

During voltage commutations, the motor operating current I_M certainly changes its slope from decreasing into increasing, but the change of magnitude is very little in comparison to changes of FC output voltage V_{Out} and HF transient of CM output current $I_{CM,Out}$. An example of waveform of motor current ripples during the whole period of controlled output current is presented in Figure 5.1. This example also confirms that motor operating currents can be assumed as unchanged within time interval of commutation process of the FC output voltage.

Second, duration of CM currents ringing after voltage commutation process is usually shorter than the modulation period. This means that CM current transients are successfully dumped before the next commutation process. Therefore, in most cases, CM currents fast transients can be analyzed as repeated single transients with steady-state periods between them long enough that they do not influence the following transient significantly.

This characteristic feature of CM transients allows assuming that during short time of commutation, motor operating currents at the output side of FC remain almost unchanged. Thus, equivalent circuit representing only single current loop for a single CM transient generated in one of phases can be extracted from the whole ASD. By means of such highly simplified circuit model,

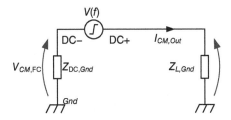

Figure 5.9 Single-circuit representation of the generation loop of the output CM current transients.

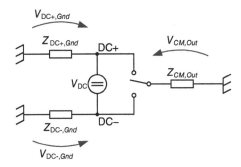

Figure 5.10 Unbalancing of CM impedances of DC buses by reconnecting CM impedance of FC load $Z_{CM,Out}$.

single commutation process can be analyzed separately. This simplification reduces extremely circuit complexity, because it allows discarding much of details resulting from a three-phase topology. Consequently, a CM transient current loop is reduced to only voltage spectrum source $V(f)$, equivalent impedance of two DC buses to ground $Z_{DC,Gnd}$, and CM impedance of load $Z_{L,Gnd}$ (Figure 5.9).

These impedances are predominantly capacitive, nevertheless, with rather complex broadband characteristics in the upper frequency range due to distributed character of partial parasitic capacitances and mutual inductances between them. Nevertheless, identification of these impedances is much easier than in three-phase configuration, as they can be measured as one-port network by using externally accessible terminals of the converter and load only. With the use of single-circuit representation, it can be also explained that output voltage switching not only generates CM voltage $V_{CM,Out}$ at the output side of FC but also CM voltage $V_{CM,FC}$ occurs internally in FC, which can be especially associated with the DC bus (Figure 5.9). The concept of CM voltage generation at the DC bus of FC can be clarified using the equivalent circuit presented in Figure 5.10.

A reconnection of the equivalent CM impedance $Z_{CM,Out}$ of FC load from one DC bus to the opposite results in an induction of unbalance between CM voltages $V_{DC+,Gnd}$ and $V_{DC-,Gnd}$ occurring between the DC buses and the ground. Therefore, the CM voltage $V_{CM,FC}$ at DC bus is defined as (5.2). In a steady state, CM voltages $V_{DC+,Gnd}$ and $V_{DC-,Gnd}$ are equal to half of V_{DC}. During commutations, these voltages change dynamically adequately to the impedance unbalance $Z_{CM,DC,Unb}$ (5.3) and therefore they generate voltage ripples of the magnitude $V_{CM,FC,PWM} = Z_{CM,DC,Unb} * V_{DC}$ correlated with modulation. The voltage ripples $V_{CM,FC,PWM}$ initiate a propagation of conducted emission generated at the output side of FC toward the power grid, whereas CM impedances of DC buses $Z_{DC+,Gnd}$ and $Z_{DC-,Gnd}$ become a fundamental coupling component between the motor side and the grid side of FC.

$$V_{CM,FC} = \frac{V_{DC+} + V_{DC-}}{2} \tag{5.2}$$

$$Z_{CM,DC,Unb} = \frac{1}{2} \left| \frac{Z_{DC-,Gnd}}{Z_{DC+,Gnd} + Z_{CM,Out}} - \frac{Z_{DC+,Gnd}}{Z_{DC-,Gnd} + Z_{CM,Out}} \right| \tag{5.3}$$

Analysis of such a strongly simplified single-circuit representation allows estimating the behavior of the entire three-phase motor side circuitry of FC, in the case when the innate unbalance of load impedances $Z_{L,Unb}$ for each phases $U1, V1, W1$ and the innate unbalance of DC bus $Z_{CM,DC,Unb}$ are significantly lower than impedance unbalance caused by the output circuitry commutations $Z_{CM,DC,Unb}$. Identification of the unbalance level of these impedances can be easily verified experimentally, as there is an external access to all terminals required for adequate measurement. A more detailed equivalent circuit for analysis of a single commutation process, taking into account dominating capacitive character of essential impedances $Z_{DC+,Gnd}$, $Z_{DC-,Gnd}$, $Z_{CM,Out,V}$ represented by adequate capacitances $C_{DC+,Gnd}$, $C_{DC-,Gnd}$, $C_{CM,V-Gnd}$, is presented in Figure 5.11.

A representation of CM currents loop by capacitances allows more comprehensive explanation of generation of PWM-related ripples of CM voltage $V_{CM,FC,PWM}$ at DC bus, presented in Figure 4.6. According to CM current loop $I_{CM,DC-}$ presented in Figure 5.11, during commutation process in phase V executed by switching-off T_{b-} and switching-on T_{b+}, equivalent capacitance of FC load of this phase $C_{CM,V-Gnd}$ must be recharged from V_{DC-} voltage to V_{DC+}. This process is forced by the DC bus voltage and can be done via capacitances (intended or parasitic), existing between the DC buses and ground $C_{DC+,Gnd}$ and $C_{DC-,Gnd}$. As it was already presented in the this chapter, during short time interval of voltage commutation, the motor operating current

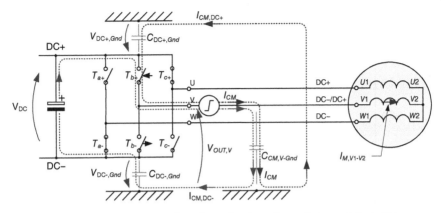

Figure 5.11 Capacitive representation of coupling impedances in loop of single CM current transient generation during commutation process at one of FC terminals.

$I_{M,V1-V2}$ can be assumed constant; thus, each single transient accompanying the reloading current can be analyzed separately as it is sufficiently distant in time from other transients.

A single transient of CM current I_{CM} as a sum of $I_{CM,DC+}$ and $I_{CM,DC-}$ is supported first by the DC bus voltage V_{DC} and second by the voltage $V_{DC+,Gnd}$ present at previously charged capacitance between the positive DC bus and ground $C_{DC+,Gnd}$. These two subcomponents of the total CM phase current $I_{CM,V}$ usually have similar values, proportional to the DC bus to ground capacitances $C_{DC-,Gnd}$ and $C_{DC+,Gnd}$, and thus are equal for DC bus with fully balanced ground capacitances. A great advantage of such a simplification is that the equivalent parasitic CM capacitances of FC load $C_{CM,V-Gnd}$ and FC DC bus in relation to ground $C_{DC+,Gnd}$ and $C_{DC-,Gnd}$ are easier to identify even within a wide frequency range in which their distributed character becomes significant and more advanced circuit models of impedance—frequency characteristics are necessary.

The elimination of an increased complexity resulting from the three-phase circuitry allows easier wide-band identification of required impedances represented as one-port networks in which one of the terminals is grounded. Ground-referenced CM impedances of FC, DC buses and FC load are less dependent on the drive operating currents as differential impedances; therefore, their experimental identification is easier because it can be performed in an offline mode when the ASD is not energized. Exemplary results of identifications of wide-band characteristics of discussed impedances are presented in Chapter 6.

An example of DC bus CM voltage changes generated during fast and slow output voltage switching is presented in Figure 5.12. The shown waveforms have been recorded during the nonconducting state of the input-side diode rectifier bridge. Between step changes generated during the transistors switchings, the

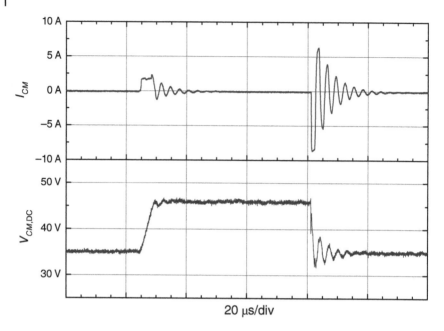

Figure 5.12 CM voltage ripples at FC DC bus during fast and slow voltage switching correlated to CM current transients.

CM voltage remains almost unchanged because the CM impedance seen at the input side of FC is high (Z_{Gnd} in Figure 5.7). During the conducting state of the input rectifier bridge–while the diodes in the conducting state significantly decrease the input-side CM impedance seen from the point of view of the FC– the DC bus CM voltage step changes correlated to the PWM are discharged via relatively low CM impedance of the power grid. Adequate waveforms are presented in Chapter 4 (Figure 4.10).

5.5 Single Transient Method

CM current transients are initiated by the voltage switching at FC output that generates voltage spectrum. The shape of the spectrum depends on the shape of the switching waveform, but mainly on the switching time. Nevertheless, spectral content of the generated CM currents may be significantly modified by CM impedances existing in ASDs. In commonly applied ASDs, CM impedance– frequency characteristics can vary significantly because of parasitic resonances that usually occur in the frequency range of conducted emission below 30 MHz.

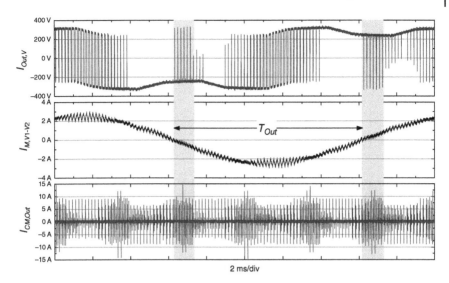

Figure 5.13 Variations of magnitudes of CM current transients during one period of the motor operating current.

Detailed parameters of each CM current transient during voltage switching can be different and they depend on many factors, as it is described in section 5.3.3. Nevertheless, an overview of typical voltage switchings during the entire period of the motor current controlled by FC allows more precise characterization of variability of CM current transient in terms of magnitudes and frequency of repetition. As the parameters of the commutation processes depend on the commutated operating currents of the motor, it should be expected that some kind of repeatability correlated with the adjusted output frequency exists. A series of CM current transients $I_{CM,Out}$, correlated with the switched voltage $V_{Out,V}$ at one of the FC outputs, recorded during the entire period T_{Out} of the motor operating current $I_{M,V1-V2}$, is presented in Figure 5.13.

First, it can be noticed that the highest magnitudes of CM current transients are repeated six times during one period of the output current. In Figure 5.13, two of the six time intervals in which peak magnitudes of CM current achieve significantly higher values than in other time intervals are highlighted. It is an effect of topological configuration of the output of three-phase transistor bridge that results in the fact that most of commutation processes (like maximum current, zero-crossing current) are repeated twice in each period in all three phases. Therefore, the frequency of those recurring changes is six times higher than the FC output frequency f_{Out}.

Second, the highest CM current transients are tightly correlated with crossing the zero value by the motor operating currents in each phase. These time intervals, from the point of view of the modulation, are associated with the

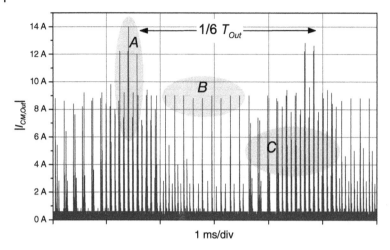

Figure 5.14 Absolute values of CM currents magnitudes during one-sixth of the motor operating current period.

use of passive vectors V_0 "000" and V_7 "111" (described in Chapter 3) when controlled motor operating currents are very low during crossing of the zero value. This correlation can also be observed on waveforms of modulated output voltage in one of the phases $V_{Out,V}$, where only limited number of relatively short switching-ons are visible, in relation to switching-on time intervals where maximum currents are controlled.

Third, it can be noticed that some of the regularities of fluctuations of the maximum magnitudes of the CM current $I_{CM,Out}$ related to the polarity of these magnitudes are correlated with polarities of operating currents in corresponding phase. Elimination of these polarity effects by focusing only on absolute values of maximum magnitudes of CM current transient helps to notice and clarify more regularities. Directions of the CM currents are usually not crucial for interfering effectiveness of the generated conducted emission as long as the summation effect of the conducted emissions generated by many sources is not present. Moreover, direction of CM currents must be balanced naturally over the time, because they flow through capacitive impedances where average values over a long period of time should be very close to zero. An exemplary waveform of absolute values for one sixth of time period of the output current is presented in Figure 5.14.

Analysis of the distribution of maximum absolute values of CM current transients over the period of FC output frequency allows further conclusions. In mid-intervals of time, between the highest magnitudes occurring in every one-sixth of the output period (in Figure 5.14 highlighted as the area *B*), absolute peak values of CM current are quite regular in magnitudes. Every second pulse

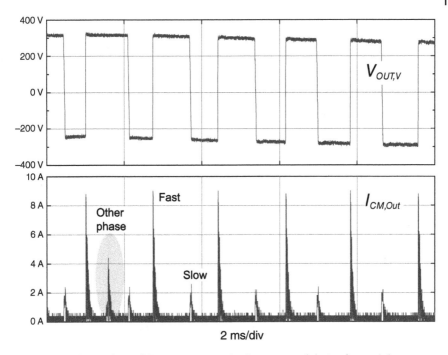

Figure 5.15 Comparison of CM currents magnitudes generated during fast and slow switchings of the converter's output voltage.

is much lower, usually approximately lower than one-third of the magnitude of regular pulses. This uniformity is correlated to two characteristic categories of output bridge switchings, fast and slow, as already described in Section 3.4.1.

A closer view on regularity of magnitudes of CM current transients is presented in Figure 5.15. Uniformity of magnitudes of CM current transients generated during fast switchings is relatively high, whereas magnitudes of transients associated with slow switchings are less regular and depend on instantaneous values of the commutated operating output currents. It can be noticed that the CM current transients can also be caused by switchings in the other phases, for example, the pulse highlighted in Figure 5.15 that appears not synchronously with the switched voltage in the observed phase V. Generally, lower CM transients are less regular but also less meaningful for the resultant levels of conducted emission, which is determined, according to standards, based on peak values occurring in the given time interval.

The time interval marked in Figure 5.14 as the area C can be characterized as a region where passive vectors are used in the modulation pattern, thus repetition of switchings and magnitudes of generated CM transient currents increase, what can be more clearly seen on zoomed waveforms of CM voltage

(a)   (b)

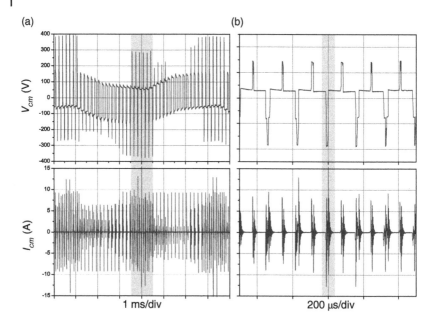

Figure 5.16 Increased CM current transient magnitudes due to simultaneous fast voltage switchings in two outputs of FC.

and current presented in Figure 5.16. The uniformity of the highest magnitudes of the generated CM current transients decreases in those time periods, due to the more complex modulation principles and possible occurrence of switchings of doubled frequency. The decreased uniformity is also an effect of shorter time interval between subsequent switchings that results in the fact that CM transient pulses do not always fully achieve steady state before the next switching. The overlap of CM current transients, depending on phase shift, can magnify or reduce the resultant maximum magnitude originated from two transients very close in time.

The overlap of CM current transients, during time intervals where passive vectors are used, in the utmost case can also lead to doubled magnitudes of resultant CM current transients, when voltage switchings occur at almost the same time in two branches of the output bridge. Such switching coincidences occur rather rarely, usually only one or few times per each one-sixth of the output period, but are consistently visible at CM currents waveforms. In Figure 5.14, this category of the highest magnitudes of CM transient currents is marked as the area A. In other words, switching of two branches of the output bridge at the same time means that voltage slope generated at the output side of FC changes by $2V_{DC}/3$ instead of $V_{DC}/3$, as it is in majority of other cases. An example of FC output-side CM voltage waveforms and the associated doubled peak values of generated CM current during almost synchronous voltage switching in two phases are presented on the right-hand side in Figure 5.16.

Summarizing, three essential characteristic categories of CM current transients $I_{CM,Out}$ generated at the FC output can be recognized from the point of view of their peak magnitudes:

- *Regular:* Accompanying fast voltage switching in only one output phase of FC at the same time, which changes converter's output-side CM voltage by one-third of V_{DC}.
- *Decreased:* Accompanying slow voltage switching in only one output phase of FC at the same time, and associated with not fully finalized voltage switching during which achieved change of switched voltage is lower than one-third of V_{DC}.
- *Elevated:* Generated due to two fast voltage switchings happening in two of the FC outputs within time interval shorter than the duration of transistor switching, which, in the utmost case, can change CM voltage at the converter's output by two-thirds of V_{DC} at once.

The most meaningful, from the point of view of conducted emission generation, are regular transients as they occur most often, approximately as often as decreased transients originated from slow switchings. Nevertheless, contribution of the decreased transients category to the overall generated conducted emission is usually not significant, as magnitudes of these transients are typically several times lower. The impact of the elevated transients on the overall conducted emission levels may vary, depending on the value of the output frequency of FC and the modulation pattern used. While the magnitudes of the elevated transients can be even two times higher in comparison to regular transients, frequency of their repetition is significantly lower than transients with regular magnitudes.

5.5.1 CM Current Transients Distribution in Individual Outputs of FC

In many research reports concerning analysis of CM currents in ASDs, only overall CM currents of FC and AC motor are considered [24,25]. Decomposition of the overall CM current of FC into separate subcomponents initiated in each phase allows revealing the principles of their generation. In Figure 5.17, waveforms of CM current transients recorded individually in each output phase of FC and correlated to overall CM current measured at the FC output are presented.

On the basis of the presented waveforms, it can be noticed that the repetition sequence and peak magnitudes of CM current transients generated in each phase $I_{CM,U}$, $I_{CM,V}$, $I_{CM,W}$ are very similar to each other and more regular in magnitudes than the resultant total CM current of FC $I_{CM,Out}$. Similar repetition sequences are shifted by 120° between each phase, adequately to

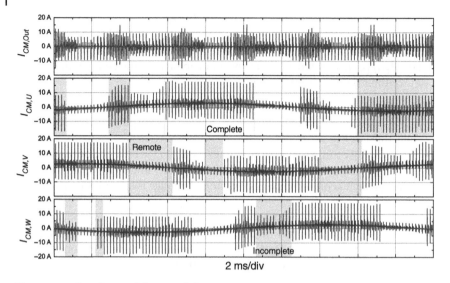

Figure 5.17 Correlation of the overall CM current $I_{CM,Out}$ measured at the FC output side with CM current transients generated in each individual phase during one period of the controlled output frequency.

motor operation currents. Three characteristic ranges of magnitudes of most often occurring CM transient in a singular output phase of FC can be visually recognized. These categories are weakly dependent on the particular output phase, but can be tightly associated with transistor switching types and defined as follows:

- *Type 1, Complete:* Associated with complete switching process of a transistor during which the voltage slope generated at the commutated output of FC reach magnitude of full V_{DC}. Adequate time periods with such transients are highlighted at the CM current waveform in phase U, $I_{CM,U}$ presented in Figure 5.17. Every second complete voltage switchings are alternately fast and slow most often. In spite of the fact that the voltage slopes generated during slow voltage switchings are as high as for fast switchings, accompanying CM current transients are much lower because of longer switching times. An example of time interval with clearly visible repetitions of slow and fast switchings is presented in Figure 5.18.

- *Type 2, Incomplete:* Associated with incomplete switching processes of a transistor during which voltage slope generated at the commutated output of the FC does not achieve magnitude of full V_{DC}. Adequate time periods with such transients are highlighted at CM current waveform in phase W, $I_{CM,W}$ in Figure 5.17. An example of voltage waveforms during incomplete voltage switching is presented in Figure 5.19.

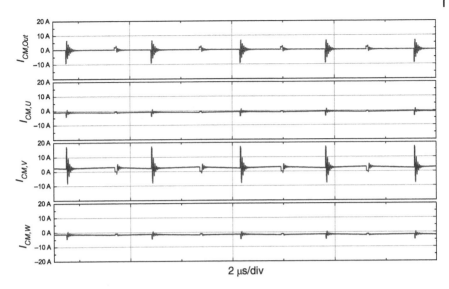

Figure 5.18 Comparison of the overall CM current $I_{CM,Out}$ measured at the FC output side with CM current transients generated in each phase individually.

- *Type 3, Remote:* Associated with voltage switching in an output phase of FC other than observed. Adequate time periods with such transients are highlighted CM current waveform in phase V, $I_{CM,V}$ in Figure 5.17. Magnitudes of these CM current transients are usually about four to six times lower than in the originating commutated phase. These currents have an opposite direction; thus, magnitudes of total CM currents of FC $I_{CM,Out}$ are smaller than magnitudes of CM current transients in individual phases $I_{CM,U}$, $I_{CM,V}$, and $I_{CM,W}$ (Figure 5.20).

5.5.2 Collection of CM Current Transients of Individual Phases into Total CM Output Current

The CM current transients generated individually in all phases are collected in the output CM current of FC. Examples of output CM current transient collection rules are presented in Figure 5.21. The left-hand side transient generated in phase V, $I_{CM,V}$, is only partly transformed into CM current of the FC $I_{CM,Out}$ because the remaining parts flowing in phases U and W ($I_{CM,U}$, $I_{CM,W}$) flow in the opposite direction in relation to $I_{CM,V}$, thus they are reducing total CM current observed at the FC output as $I_{CM,Out}$. The right-hand side transients generated in short time distance in two phases, U and W ($I_{CM,U}$, $I_{CM,W}$), are also only partially cumulated into the total CM current observed in the FC output, because return currents initiated by these two phases are flowing in the

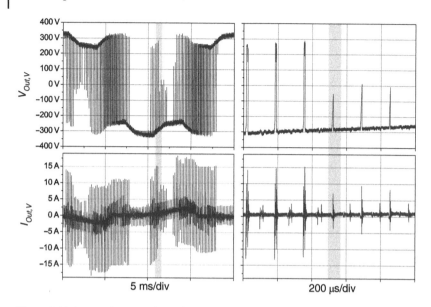

Figure 5.19 Incomplete switchings of voltage at FC outputs during short switching-on sequences.

opposite direction in the third phase V. Nevertheless, output voltage switching in two phases within close time distance between them results in significant increase of magnitude of the resultant total CM current observed in the FC output $I_{CM,Out}$ in relation to the previous transient.

The resultant CM current is difficult to predict precisely. Nevertheless, some helpful simplification can be introduced for accessible estimation of resultant CM currents. Representative, from the point of view of phase CM transient collection into total CM current of FC, subintervals of the entire modulation period of the output operating currents are indicated in Figure 5.22. The same waveforms that were previously analyzed from the point of view of magnitudes of CM transients in individual phases (Figure 5.17) here are correlated to different categories of collection techniques.

According to the characteristic time coincidences between the subcomponents of CM transients generated in each phase, a number of distinguishable intervals have been denoted and marked as A, B, C, D, E, and F (Figure 5.22):

- *Interval A:* The most clear case during the entire period of the controlled output current, where complete switchings take place only in one phase and initiate CM current transients $I_{CM,V}$ of regular magnitudes. Smaller part of these transients returns back to the converter as DM current transients via two other phases $I_{CM,U}$ and $I_{CM,W}$, and the remaining part via grounding connection forming FC's CM current $I_{CM,Out}$ (as explained in Section 3.4.2).

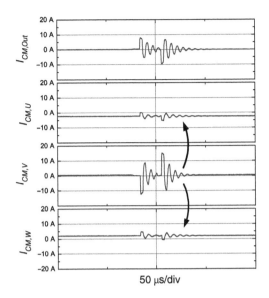

Figure 5.20 Remote type of a CM current transient generated in other two phases $I_{CM,U}$, $I_{CM,W}$, by CM current transient in the commutated phase $I_{CM,V}$.

Due to high magnitudes of the controlled motor operating current, during these intervals the duty cycle in switched phase is close to 50%; therefore, generated CM transients are very regular, alternately high, and low adequately to fast and slow switchings.

- *Interval B:* Output voltage switchings occur in two phases; in one of the commutated phases, the switchings are complete $I_{CM,V}$; in the second, they are incomplete $I_{CM,W}$. In the third phase, only the return DM current transients flow back to the converter. As the larger parts of these two current transients originated from two commutated phases $I_{CM,V}$ and $I_{CM,W}$ are collected into the CM current of FC, the waveforms of the resultant CM current transients $I_{CM,Out}$ become more complex. Repetitions of CM transients are about being doubled in average and summing effects noticeably depend on the time shifts between the following transients. Nevertheless, dominating peak magnitudes are mainly associated with complete switchings in phase *W* and very similar in magnitudes to those occurring in the interval *A* in which only one phase was commutated.
- *Interval C:* Similar to interval *B*, two phases are commutated, but both in complete switching mode. Therefore, repetition frequency of CM current transients $I_{CM,Out}$ is doubled, but peak magnitudes remain at similar level, analogously as during intervals *A* and *B*.
- *Interval D:* All three phases are commutated, two in complete switching mode and the third one in incomplete switching mode. Because of the com-

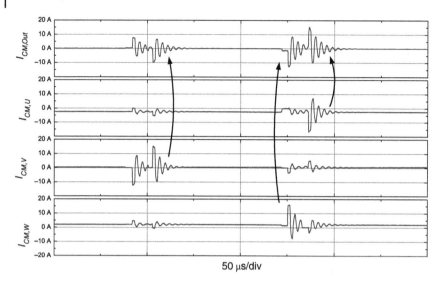

Figure 5.21 Secondary CM current transients generated in other two phases $I_{CM,U}$, $I_{CM,W}$, by CM current transient in the commutated phase $I_{CM,V}n$.

mutations, in all three phases, the repetition frequency of CM transients $I_{CM,Out}$ increases further, up to triple the frequency in relation to interval A.

- *Interval E*: All three phases are commutated, all in complete switching mode. The resultant magnitudes of CM current $I_{CM,Out}$ are only slightly different in relation to the interval D. Observed peak magnitudes should be more regular due to regular switchings in complete mode in all three phases. Nevertheless, due to additional influences, it is difficult to notice the difference in the regularity of magnitudes, because incomplete switchings in one of the three commutated phases, underlined in description of interval D occur very rarely.

- *Interval F*: This is a very specific case, which can occur during commutations in all three phases (thus, in the intervals C, D, or E), when switchings in two phases occur in almost the same time, with delay shorter than the fast switching time. Such coincidence usually occurs only a few times during one-sixth of the output operating current period, while one of the motor operating phase currents is crossing the zero value.

It should be underlined that the presented categorization of commutation modes is exemplary. Therefore, the duration of the discussed characteristic time subintervals correlated with modulation pattern can differ significantly depending on many parameters, primarily on the motor load. However, it has been verified experimentally that the described processes of generation converter's CM current transients are very representative and occur in vast majority of operation modes of ASDs. Magnitudes of CM transients generated

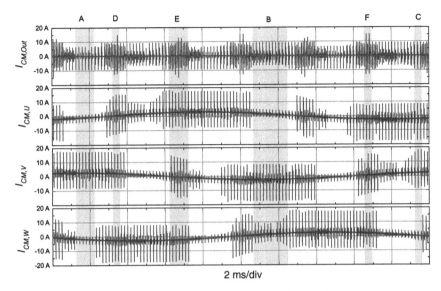

Figure 5.22 Characteristic subintervals during the period of controlled motor current meaningful for generated overall CM currents of FC.

in individual phases are only slightly dependent on variations of the motor operating currents. Nevertheless, due to differences in duration of particular subintervals, the resultant CM current of FC can vary noticeably with variation of the motor load. Taking working conditions of commonly used ASDs into account (which are usually significantly loaded), some of the defined intervals are more possible to occur than others. During a typical operation of an ASD, intervals of the type *B*, *C*, *D*, *E* are used for the vast majority of drive operation time, whereas the intervals *A* and *F* occur less often.

5.6 Summary

The presented considerations and analysis based on experimentally obtained data reveal that magnitudes of CM currents' transients originated in individual phases are more regular than the resultant total CM current of FC, especially during time intervals where complete commutation processes take place. Existing irregularities are correlated with the defined characteristic subintervals of periods of the controlled output operating currents. Therefore, if output voltage switchings happen in the given time in which particular category takes place, magnitudes of generated CM transients are similar; so they are possible to predict by analyzing exemplary cases of transients only.

The resultant CM current of FC, which is an effect of switching processes in all individual phases, is more irregular than CM current transients in particular phases, because it depends additionally on many interactions between phases. Especially the overlap of CM transient currents generated in all three phases results in increase of variability of peak magnitudes and repetition frequencies. Peak magnitudes of the CM current of FC are usually about 20–30% lower than in individual phases, because part of CM transients generated in each phase is returning back to FC via DM parasitic capacitances. Nevertheless, it is possible to determine the most frequently occurring magnitudes of FC CM transients that are quite similar, except for relatively rarely occurring case, where occurring magnitudes can be up to doubled due to the possible incidental synchronous switchings in two phases.

From the point of view of interfering potential of nearby devices by generated CM transients, the peak magnitudes are the most meaningful; thus, higher values observed in individual phases can be most significant for interfering devices located close to the output side of FC. Nevertheless, peak magnitudes of CM currents of FC, which are statistically lower but more varying than phase transients in individual phases, can be transferred toward the power grid via DC bus of FC. For majority of cases, this affects the overall levels of conducted emission at the input side of FC that are limited by standard recommendations.

Despite high variability of many factors influencing CM current transients generation at the FC output side, it was demonstrated that

- the highest and most frequently occurring transients generated during fast switchings are similar in magnitudes,
- these transients are not significantly dependent on the motor operating currents and mostly occur as single transients in separated time intervals.

Therefore, spectral content of typical single transients generated during characteristic time intervals is expected to be representative for estimation of the overall level of conducted emission of an ASD without the necessity of detailed wide-band modeling of the entire ASD system. Exemplary verifications for this statement are presented in the following chapters after presentation of the methods of wide-band identifications of adequate CM impedances.

The presented approach allows brief and effective estimation of expected spectral density of CM currents generated at the output side of FC, which are the most meaningful for determination of conducted emission of ASD. This can be achieved by analyzing only the representative switching transients using highly simplified single-loop circuit model, which significantly reduces difficulties related to broadband identification of fundamental ASD components. Necessary for applying this method, broadband characteristics of only limited number of selected CM impedances of ASD components are possible to identify based on externally accessible measurement only, with relatively low experimental overheads.

References

1 Q. Liu, W. Shen, F. Wang, D. Boroyevich, V. Stefanovic, and M. Arpilliere, "Experimental evaluation of IGBTs for characterizing and modeling conducted EMI emission in PWM inverters," in *2003 IEEE 34th Annual Power Electronics Specialist Conference, 2003 (PESC'03)*, vol. 4, IEEE, 2003, pp. 1951–1956. Available at http://ieeexplore.ieee.org/xpls/abs_all.jsp?arnumber=1217751

2 Z. Wang, X. Shi, Y. Xue, L. Tolbert, and B. Blalock, "A gate drive circuit of high power IGBTs for improved turn-on characteristics under hard switching conditions," in *2012 IEEE 13th Workshop on Control and Modeling for Power Electronics (COMPEL)*, 2012, pp. 1–7.

3 H. Zhu, J.-S. Lai, J. Hefner, A.R., Y. Tang, and C. Chen, "Modeling-based examination of conducted EMI emissions from hard and soft-switching PWM inverters," *IEEE Transactions on Industry Applications*, vol. 37, no. 5, pp. 1383–1393, Sept./Oct. 2001.

4 R. J. Kerkman, D. Leggate, D. W. Schlegel, and C. Winterhalter, "Effects of parasitics on the control of voltage source inverters," *IEEE Transactions on Power Electronics*, vol. 18, no. 1, pp. 140–150, Jan. 2003.

5 J. Luszcz, "Broadband modeling of motor cable impact on common mode currents in VFD," in *2011 IEEE International Symposium on Industrial Electronics (ISIE)*, IEEE, 2011, pp. 538–543. Available at http://ieeexplore.ieee.org/xpls/abs_all.jsp?arnumber=5984215

6 J.-S. Lai, X. Huang, E. Pepa, S. Chen, and T. Nehl, "Inverter EMI modeling and simulation methodologies, " *IEEE Transactions on Industrial Electronics*, vol. 53, no. 3, pp. 736–744, June 2006.

7 J. Luszcz, "AC motor feeding cable consequences on EMC performance of ASD," in *2013 IEEE International Symposium on Electromagnetic Compatibility (EMC)*, 2013, pp. 248–252. Available at http://ieeexplore.ieee.org/stamp/stamp.jsp?arnumber=6670418

8 Q. Liu, F. Wang, and D. Boroyevich, "Conducted-EMI prediction for ac converter systems using an equivalent modular 2013; terminal 2013; behavioral (MTB) source model," *IEEE Transactions on Industry Applications*, vol. 43, no. 5, pp. 1360–1370, Sept.–Oct. 2007.

9 M. Moreau, N. Idir, P. L. Moigne, and J. J. Franchaud, "Utilization of a behavioural model of motor drive systems to predict the conducted emissions," in *2008 IEEE Power Electronics Specialists Conference*, June 2008, pp. 4387–4391.

10 D. Gonzalez, J. Gago, and J. Balcells, "New simplified method for the simulation of conducted EMI generated by switched power converters," *IEEE Transactions on Industrial Electronics*, vol. 50, no. 6, pp. 1078–1084, Dec. 2003.

11 C. Jettanasen, F. Costa, and C. Vollaire, "Common-mode emissions measurements and simulation in variable-speed drive systems, " *IEEE Transac-*

tions on Power Electronics, vol. 24, no. 11, pp. 2456–2464, 2009. Available at http://ieeexplore.ieee.org/xpls/abs_all.jsp?arnumber=5346844

12 P. Musznicki, J.-L. Schanen, P. Granjon, and P. Chrzan, "The Wiener filter applied to EMI decomposition," *IEEE Transactions on Power Electronics*, vol. 23, no. 6, pp. 3088–3093, Nov. 2008.

13 J. Adabi, F. Zare, A. Ghosh, and R. Lorenz, "Calculations of capacitive couplings in induction generators to analyse shaft voltage," *Power Electronics*, IET, vol. 3, no. 3, pp. 379–390, May 2010.

14 J.-S. Lai, X. Huang, S. Chen, and T. W. Nehl, "EMI characterization and simulation with parasitic models for a low-voltage high-current ac motor drive, " *IEEE Transactions on Industry Applications*, vol. 40, no. 1, pp. 178–185, 2004. Available at http://ieeexplore.ieee.org/xpls/abs_all.jsp?arnumber=1268194

15 H. Ke, T. Hubing, and F. Maradei, "Full-wave electromagnetic modelling from dc to GHz using FEM-spice," in *2010 Asia-Pacific International Symposium on Electromagnetic Compatibility*, April 2010, pp. 508–511.

16 N. Djukic, L. Encica, and J. J. H. Paulides, "Electrical machines: comparison of existing analytical models and FEM for calculation of turn-to-turn capacitance in formed windings," in *2016 Eleventh International Conference on Ecological Vehicles and Renewable Energies (EVER)*, April 2016, pp. 1–8.

17 I. F. Kovacevic, T. Friedli, A. M. Musing, and J. W. Kolar, "3-D electromagnetic modeling of parasitics and mutual coupling in EMI filters," *IEEE Transactions on Power Electronics*, vol. 29, no. 1, pp. 135–149, 2014. Available at http://ieeexplore.ieee.org/stamp/stamp.jsp?arnumber=6484987

18 I. F. Kovacevic, T. Friedli, A. M. M using, and J. W. Kolar, "Full PEEC modeling of EMI filter inductors in the frequency domain," *IEEE Transactions on Magnetics*, vol. 49, no. 10, pp. 5248–5256, Oct. 2013.

19 M. Degano, P. Zanchetta, L. Empringham, E. Lavopa, and J. Clare, "HF induction motor modeling using automated experimental impedance measurement matching," *IEEE Transactions on Industrial Electronics*, vol. 59, no. 10, pp. 3789–3796, Oct. 2012.

20 J. Sun and L. Xing, "Parameterization of three-phase electric machine models for EMI simulation," *IEEE Transactions on Power Electronics*, vol. 29, no. 1, pp. 36–41, Jan. 2014.

21 K. Jia, G. Bohlin, M. Enohnyaket, and R. Thottappillil, "Modelling an ac motor with high accuracy in a wide frequency range," *IET Electric Power Applications*, vol. 7, no. 2, pp. 116–122, Feb. 2013.

22 I. Stevanovi, B. Wunsch, G. L. Madonna, and S. Skibin, "High-frequency behavioral multiconductor cable modeling for EMI simulations in power electronics," *IEEE Transactions on Industrial Informatics*, vol. 10, no. 2, pp. 1392–1400, May 2014.

23 M. Shahjalal, H. Lu, and C. Bailey, "A review of the computer based simulation of electro-thermal design of power electronics devices," in *20th International Workshop on Thermal Investigations of ICs and Systems*, Sept. 2014, pp. 1–6.

24 F. Costa, C. Gautier, E. Laboure, B. Revol, F. Costa, C. Gautier, E. Laboure, and B. Revol, *Electromagnetic Compatibility in Power Electronics.* John Wiley & Sons, Inc., New York, 2013. Available at http://dx.doi.org/10.1002/9781118863183.ch1

25 S. Wang, Y. Y. Maillet, F. Wang, R. Lai, F. Luo, and D. Boroyevich, "Parasitic effects of grounding paths on common-mode EMI filter's performance in power electronics systems," *IEEE Transactions on Industrial Electronics*, vol. 57, no. 9, pp. 3050–3059, Sept. 2010.

6

Broadband Behavior of Fundamental Components of ASD

The ability to simplify means to eliminate the unnecessary so that the necessary may speak.

Hans Hofmann

The analysis of generation and propagation of conducted emission in ASD applications requires the recognition of impedance characteristics of all essential subcomponents within adequately wide-frequency range. Conducted emission is defined by standards as spectral characteristic of CM voltages measured at input terminals of devices connected to the power grid. Therefore, particularly meaningful are CM impedances that in ASD mainly depend on parasitic capacitive couplings of drive components in relation to the ground. At the stage of application of ASDs to the FC—as already assembled and fixed device—external components are connected. These external components are individually selected to fulfill ASD application requirements and can differ significantly, especially in terms of broadband behavior that is usually not sufficiently specified by manufactures. Therefore, identification of broadband impedance characteristics of external ASD components is a crucial preparative stage for analysis of conducted emission.

6.1 Behavior of Real-World Passive Devices within Frequency Range of Conducted Emission

In the frequency range of conducted emission (i.e., 30 MHz), a vast majority of passive components used in power circuits in ASD applications expose

High Frequency Conducted Emission in AC Motor Drives Fed by Frequency Converters: Sources and Propagation Paths, First Edition. Jaroslaw Luszcz.
© 2018 by The Institute of Electrical and Electronic Engineers, Inc. Published 2018 by John Wiley & Sons, Inc.

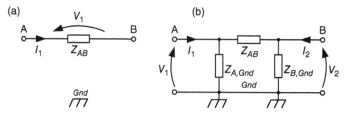

Figure 6.1 Representation of a component with two terminals for DM signals analysis as one-port network (a) and for CM signals as two-port network (b).

parasitic couplings that are meaningful in terms of impact on relative impedances [1–3]. Primarily, it is a result of considerable sizes of those components with high-ampere capacity and that require insulation strength, which results in parasitic capacitances and inductances difficult to reduce. Therefore, in the upper frequency range of conducted emission, commonly used components in ASDs such as capacitors, inductors, and even wire connections, behave unconventionally and exhibit not only intended features, but simultaneously a mixture of them where each one is frequency dependent [4–6]. Additionally, parasitic couplings existing between components and the ground, which are especially important in conducted emission generation and propagation, introduce an extra artificial terminal to each components—the ground terminal. This makes one-port network representation of components commonly used for analysis of DM phenomena be replaced by two-ports network models for analysis of CM phenomena (Figure 6.1).

The most meaningful external components of ASD influencing the generated conducted emission are as follows:

- AC motor as a set of windings arranged into three phase topology
- Motor feeding cable of a particular length
- FC input filters applied in order to limit the input current harmonics and/or conducted emission
- FC output filters used for AC motor protection against excessive voltage harmonics and/or winding over voltages.

All of those components are passive. AC motors with feeding cables can have significantly different parameters but usually are of similar configuration; however, the input and output filters are assembled of capacitances and inductances that can be arranged in various topologies. Therefore, broadband behavior of those categories of passive components is essential for conducted emission estimation.

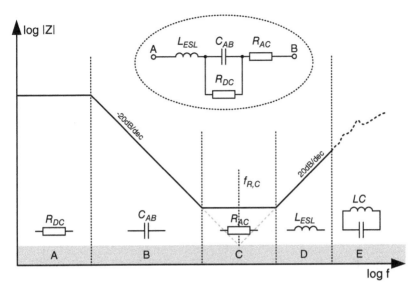

Figure 6.2 Broadband circuit model and general theoretical DM impedance characteristic of capacitors.

6.2 Impedance-Frequency Characteristics of Capacitors

Capacitors are widely used in ASD applications as subcomponents for filtering conducted emission as they decrease levels of CM and CM impedances of devices, especially in HF rage. Capacitors, as high-volume manufactured components, are usually well characterized by manufactures, also in HF range [7,8]. From the point of view of efficient filtering conducted emission, the most meaningful parameter of capacitor is parasitic inductance known as equivalent serial inductance (ESL) specified by manufactures. Parasitic inductance is an inevitable effect of internal construction of capacitors and decreases the frequency above which capacitor's behavior becomes inductive. A general impedance–frequency characteristic of typical capacitors is presented in Figure 6.2, where in the frequency sub-band marked as D, an increase of capacitor's impedance caused by ESL can be observed.

Another common way of specifying upper frequency, below which capacitors can be effective as means of HF filtering is their self-resonant frequency, defined as $f_{r,C}$ (6.1). Above the capacitor's resonance frequency, inductive character of impedance becomes predominant, thus its impedance increase, with the increases of frequency.

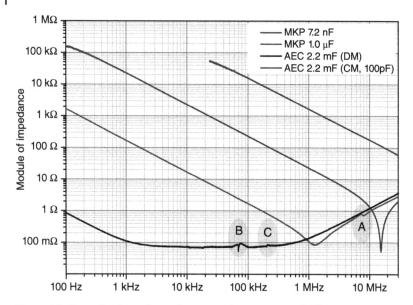

Figure 6.3 Exemplary impedance–frequency characteristics of different types of capacitors used in power electronic applications: aluminum electrolytic capacitor (AEC), metalized polypropylene film capacitors (MKP).

$$f_{r,C} = \frac{1}{2\pi\sqrt{C_{AB}L_{ESL}}} \tag{6.1}$$

Summarizing the most important parasitic parameter of capacitors, from the point of view of behavior in HF range, is its serial inductance that reduces their effectiveness for conducted emission filtering. Impedance frequency characteristic of exemplary types of capacitors that are commonly used in power electronics application where conducted emission paths occur are presented in Figure 6.3. In Figure 6.4, impedance–frequency characteristic of exemplary types of capacitors commonly used for filtering conducted emission inside power electronic converters and also in external RFI filters are presented.

Based on the presented representative impedance characteristics it can be noticed that

- all characteristics are strongly linear in logarithmic scale within the frequency range below serial resonance frequency, which means that impedance character is almost purely capacitive, the slope of impedance characteristics is −20 dB per decade,
- most of capacitors used in high-power applications exhibit characteristic resonance frequency f_{rC} below 30 MHz, thus within the frequency range of

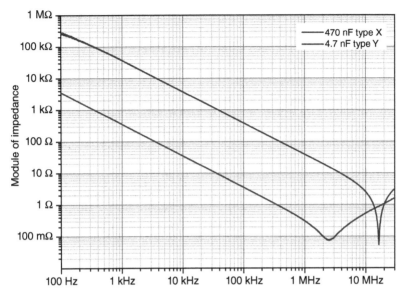

Figure 6.4 Exemplary impedance–frequency characteristics of type X and Y capacitors commonly used for filtering of conducted emission at the grid side of power electronic converters.

conducted emission, which results with the risk of reduced filtering performances usually above few megahertz,

- parasitic resonances, other then the capacitor's main resonance, are rather not meaningful, for example, capacitor MKP 1.0 μF at the frequency of about 8 MHz (Figure 6.3 A) and capacitor AEC 2.2 mF at the frequencies of about 70 and 200 kHz (Figure 6.3 B and C),
- electrolytic capacitors of high capacity can have relatively low serial inductance ESL, thus they exhibit low impedance within really broad frequency range, even up to few hundreds of kilohertz, for example, Figure 6.3 capacitor AEC 2.2 mF,
- electrolytic capacitors, despite the fact of being very bulky, exhibit very low parasitic capacitances in relation to ground, Figure 6.3 characteristic AEC 2.2 mF (CM 100pF).

Presented examples convince us that broadband modeling of capacitors used for conducted emission filtering in ASDs, as high-power electronic applications, can be successfully executed using relatively simple circuit models, such as presented in Figure 6.2, which are also often provided by manufactures.

6.3 Impedance-Frequency Characteristics of Inductors

Inductors, especially those of high rated currents that are applied in high-power applications such as ASDs, are not as massively produced as capacitors. Such inductors are usually manufactured for specific use in a particular class of ASD application, and therefore are not so thoroughly parameterized in other applications. Therefore, selection of inductors for filtering conducted emission in extraordinary applications of ASDs with special demands very often requires individual experimental identification of necessary parameters, especially in HF range [9–11].

From the point of view of their behavior in HF range, the most meaningful parameters of inductors are parasitic capacitances, which reduce inductor's impedance in the frequency range above the characteristic self-resonant frequency defined as $f_{r,L}$ (6.1). L_{AB} is inductor's equivalent inductance and C_{EPC} inductor's equivalent parallel parasitic capacitance that are determined for two terminal models.

$$f_{r,L} = \frac{1}{2\pi \sqrt{L_{AB} C_{EPC}}} \qquad (6.2)$$

A theoretical impedance frequency characteristic of inductors as one-port components is presented in Figure 6.5. According to this characteristic, inductive character of impedance correlated with equivalent inductance L_{AB} dominates in the frequency subrange B. In higher frequency range (in the subrange D), the capacitive character of impedance is dominating, which depends on parasitic equivalent capacitance C_{EPC}. Impedance of an inductor for frequencies close to the resonance frequency $f_{r,L}$ (subrange C) depends on inductor's losses specified for this frequency range, represented in a circuit model by the equivalent resistance R_{AC}.

Unfortunately, impedance characteristics of real-world high power core inductors used in ASD applications can be significantly different in relation to the theoretical impedance characteristic presented in Figure 6.5 due to some extra phenomena that increase in HF range and distributed character of parasitic capacitances. To visualize these differences, an example of full impedance characteristics, absolute value and phase, of a real-world inductor of the rated current 16 A and inductance 10 mH are presented in Figure 6.6.

First, with the increase of frequency more eddy currents are generated in the windings, the core structure, and other conductive components nearby. Increased generation of eddy currents results in decrease of inductance and increase of losses. Losses in windings represented by equivalent resistance correlated to HF AC current also increase due to skin and proximity effects that strengthen with the increase of frequency. Due to these phenomena, a slope

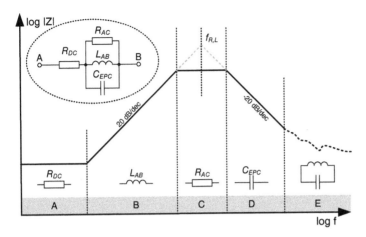

Figure 6.5 One-port broadband circuit model and general theoretical impedance frequency characteristic of inductors.

of impedance frequency characteristic in frequency subrange B (Figure 6.5) is usually lower than 20 dB per decade, which can be noticed in the measured impedance characteristic of exemplary inductor presented in Figure 6.6 in the frequency range below the main parallel resonance $f_{Rp,L}$.

Second, parasitic capacitances of high-power inductors are usually not equally distributed along winding because of windings layers and constructional sectioning. Usually, a number of extra local resonances, weaker than the main resonance $f_{Rp,L}$, occur in HF range, which are especially difficult to avoid or minimize because of bulky structures of windings. Unequal distribution of parasitic capacitances usually leads to formation of local resonance circuits that can be especially significant for resultant impedances in the frequency subranges D and E according to Figure 6.5. In the presented impedance characteristics of real-world inductor in Figure 6.6, adequate frequency range can be recognized above 2 MHz.

Third, equivalent parasitic capacitance C_{EPC} is a result of not only self-capacitance of winding, but also of CM parasitic capacitances between winding and ground that can be of significant value and usually unequally distributed in relation to both terminals. CM capacitances of inductors are particularly difficult to reduce in inductors applied in high-power applications where magnetic core and mechanical construction are of significant value, made of conductive staff and grounded. CM capacitances of inductors are especially meaningful for filtering of conducted emission, because they can substantially change CM impedances of filtered lines. Therefore, such inductors should be analyzed as two-port networks with common ground terminal, as a third terminal, according to Figure 6.1.

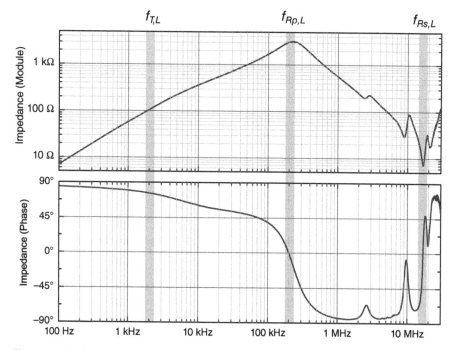

Figure 6.6 Bode plot of impedance-frequency characteristics of evaluated exemplary inductor.

6.3.1 Circuit Modeling of Frequency-Dependent Inductance of Inductors

The range of changes of inductor's fundamental parameters, the equivalent inductance L_{AB} and the resistance R_{AC}, with the increase of frequency is dependent on inductor's structure and especially winding's designs, and can be optimized within a limited range. Nevertheless, in most cases these are nonlinear phenomena that are difficult to model within a wide-frequency range. Within a limited frequency range, methods based on fractional order modeling can be successfully used [12,13], but they require laborious identification processes. Nonlinear processes occurring in inductors, mainly related to eddy currents, proximity, and skin effects, usually become significant in frequency range above a few kilohertz. Effects of these phenomena can be observed at inductor's impedance Bode diagrams, as an increase of the slope of phase characteristic and a decrease of the slope of impedance module characteristic (Figure 6.7) above the transition frequency defined as $f_{T,L}$. This transition frequency is weakly detectable and can be estimated by intersection of the tangent lines a and d (Figure 6.7). In the evaluated case it is above the frequency of about 2 kHz.

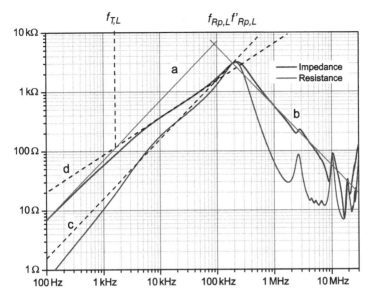

Figure 6.7 Frequency characteristics of equivalent impedance and resistance of the evaluated inductor.

In Figure 6.7, the slope of impedance characteristic close to 20 dB per decade is indicated by the line a, and it reflects well enough the impedance increase in frequency range below a few kilohertz. The line d indicates decreased slope of impedance characteristic and corresponds to the impedance frequency characteristic in the frequency range around a few tens of kilohertz. Vertical distance between the lines a and c corresponds to the inductor's inductance difference for LF (at about 100 Hz) and HF (at about 100 kHz). The line b represents the inductor's impedance correlated to the equivalent parasitic capacitance that becomes dominant for the inductor's impedance magnitudes in the frequency range above the self-resonance frequency $f_{Rp,L}$.

It should be underlined that the measured self-resonance frequency of inductors $f_{Rp,L}$ is moved toward higher frequencies in relation to $f'_{Rp,L}$ determined for the given equivalent parasitic capacitance C_{EPC} represented by the line b and inductance defined at LF represented by the line a. The difference between those self-resonance frequencies, real $f_{Rp,L}$ and modeled using simple circuit model $f'_{Rp,L}$ presented in Figure 6.7, corresponds to overall change of the inductor's inductance with frequency and cannot be properly reflected by a model with only one inductance L_{AB}.

In the frequency range below the self-resonance frequency $f_{Rp,L}$, along with deceasing impedance, a corresponding change of slope of the equivalent resistance characteristic can be observed (Figure 6.7). This effect can also be

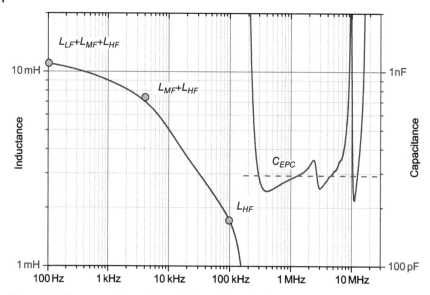

Figure 6.8 Frequency dependence of the evaluated inductor's inductance and equivalent parasitic capacitance.

more clearly seen in the inductor's impedance phase characteristic (Figure 6.6), where in the frequency range of about 10 kHz phase angle decreases more than below 1 kHz. The detailed inductance frequency characteristic of evaluated inductors is presented in Figure 6.8, where it can be noticed that inductance reduction is significant, about 80% at the frequency of 70 kHz. In the same figure, it can be seen that the equivalent parasitic capacitance, determined in the frequency range above $f_{Rp,L}$, is also slightly frequency dependent and increases with frequency.

Simultaneous and reverse frequency dependence of the inductor's inductance and resistance allow applying modeling approach based on circuit model consisting of a number of sections connected in series consisting of inductance and resistance connected in parallel (Figure 6.9). This method of realization of an equivalent circuit model is used for representing nonlinear phenomena, likewise, iron core nonlinearity in magnetic components and is known as the "first Foster circuit" that allows achieving required accuracies using the adequately high number of LC sections [12,14].

According to the Foster's circuit representation, an impedance characteristic can be formulated as (6.3), where R_0 represents inductor's equivalent resistance at LF and L_0 represents equivalent inductance at HF. Adequately, inductors equivalent impedance at LF and HF can be approximated as (6.4) and (6.5), respectively.

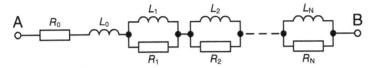

Figure 6.9 Realization of frequency-dependent inductance and resistance using the first Foster circuit model.

$$Z(s) = \left(R_0 + sL_0\right) + \sum_{i=1}^{N} \frac{R_i s L_i}{R_i + sL_i} \tag{6.3}$$

$$Z(s)_{@LF} \approx R_0 + s \sum_{i=1}^{N} L_i \tag{6.4}$$

$$Z(s)_{@HF} \approx \sum_{i=1}^{N} R_i + sL_N \tag{6.5}$$

The procedure of parameter determination of this model, especially for high number of model sections, is a typical optimization problem that require estimation of rational initial values, which can be improved by successive iteration. Unfortunately, there are many circuit realizations that can be found. Depending on the used iteration method and selection of the starting point parameters, this method can lead to well-suited results that are not well correlated with the modeled physical phenomena. Therefore, initial estimatison of iteration starting point parameters of the model and rational limitation of the number of sections is essential for success in obtaining profitable models.

Search for fulfilling starting point parameters can be started for determination of boundary inductances and resistances occurring at outer limits of the considered frequency band, taking into account the relationship between (6.4) and (6.5). Furthermore, preliminary assessment of the shape of the experimentally obtained impedance characteristic is essential for selection of reasonable number of model's sections.

An example of a three-sectional circuit model of frequency-dependent impedance characteristic is presented in Figure 6.10, for which two sets of three equations can be formulated; the first set for inductances (6.6) and the second set for resistances (6.7).

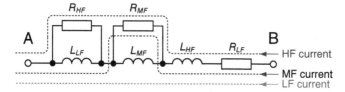

Figure 6.10 Sample of three-sectional circuit model of inductor's frequency dependent equivalent inductance and resistance.

$$L\left(f_{L1}\right) = L_{HF} + L_{MF} + L_{LF}$$
$$L\left(f_{L2}\right) = L_{HF} + L_{MF} \qquad (6.6)$$
$$L\left(f_{L3}\right) = L_{HF}$$

$$R\left(f_{L1}\right) = R_{LF}$$
$$R\left(f_{L2}\right) = R_{LF} + R_{MF} \qquad (6.7)$$
$$R\left(f_{L3}\right) = R_{LF} + R_{MF} + R_{HF}$$

These relations are helpful for finding initial parameters as close as possible to the foreseen, physically reasonable solution and they significantly limit necessary further efforts during the iteration process. First, values of resistance at low frequency $R_{LF} = R(f_{LF})$ and inductance at high frequency $L_{HF} = L(f_{HF})$ can be directly determined based on the impedance characteristic. Second, the range of changes of inductance ΔL and resistance ΔR can be calculated according to the formulas (6.8) and (6.9), respectively, as $\sum_{i=1}^{N} L_i = L(f_{LF})$ and $\sum_{i=1}^{N} R_i = R(f_{HF})$. Third, it is recommended to chose values of starting points of intermediate inductances and resistances that are logarithmically equidistantly spread within their ranges of change ΔL and ΔR.

$$\Delta L = \sum_{i=1}^{N} L_i - L_1 = L(f_{LF}) - L_{HF} \qquad (6.8)$$

$$\Delta R = \sum_{i=1}^{N} R_i - R_1 = R\left(f_{HF}\right) - R_{HF} \qquad (6.9)$$

General idea of the use of the Foster model is that with the increase of frequency impedances associated with inductances in each sections also increase, thus for frequencies higher than the characteristic transition frequency $f_T i = R_i / L_i$ of particular section, greater part of current flows via resistances.

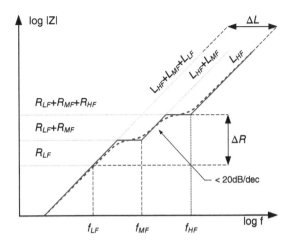

Figure 6.11 Graphical representation of simplified method of modeling frequency-dependent inductances.

Therefore, the increase of equivalent resistance and the decrease of inductance is observed with the increase of frequency, which is graphically presented in Figure 6.10 as HF and LF current paths. More beneficial explanation can be visualized with the use of simplified theoretical impedance characteristics presented in Figure 6.11.

According to the presented simplified impedance characteristics for circuit model with three sections, the resultant theoretical impedance characteristic becomes a polyline. Nevertheless, in case when characteristic frequencies of nearby sections f_{Ln} and $f_{L(n+1)}$ are closer than one decade in frequency domain, the resultant characteristic in practice becomes quite smooth. To obtain reasonably accurate models of changing inductance of inductors commonly used in power electronic applications, usually the use of only one or two sections per each decade is enough to achieve satisfactory adequacy within the analyzed frequency range.

6.3.2 Circuit Modeling of Distributed Parasitic Capacitances

Distributed character of partial parasitic capacitances of inductor's windings results in that their representation by only one equivalent capacitance C_{EPC} (Figure 6.5) is a simplification that can be accepted only up to a limited frequency. The main self-resonance frequency of inductors $f_{Rp,L}$ in most cases used in power electronic applications is very pronounced, which is also visible on the impedance characteristic of the evaluated inductor in Figure 6.6.

The dominant character of the main self-resonance frequency allows extending the range of frequency of successful applicability of the discussed circuit

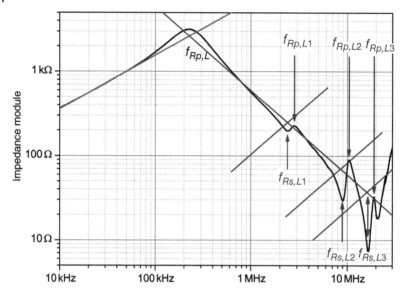

Figure 6.12 Characteristic resonance frequencies originated by distributed partial parasitic capacitances in the frequency range above the inductor's main self-resonance $f_{Rp,L}$.

model somewhat above the self-resonance frequency $f_{Rp,L}$—usually for about one decade. Nevertheless, for most of the commonly used inductors, impedance characteristics in higher frequencies become influenced by distributed character of parasitic capacitances, thus they are more complex and can not be adequately reflected by single equivalent capacitance C_{EPC}. In the evaluated inductor, these effects can be noticed above frequency of about 1 MHz where extra resonances occur (Figure 6.12) between 2 and 3 MHz, about 10 MHz.

To adequately represent inductor's internal resonances caused by distributed character of parasitic capacitances, a circuit model must be expanded by additional nodes that allow to reflect these resonances.

Detailed mapping of a physical structure of partial parasitic capacitances connection topology requires to use more advanced theoretical methods like the finite element method (FEM) [15–17]. Other method that can be used is the vector fitting (VF) approach, which allows for accurate impedance representation of one-port network by mathematical formulas [18,19]. Unfortunately, results obtained by using these methods, despite the high accuracy, do not provide efficient possibilities for formulation of simplified circuit representation with reasonable level of complexity correlated to the most meaningful resonances.

Closer look at many of experimentally obtained impedance characteristics of inductors commonly used in high-power applications allows for some gen-

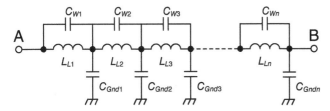

Figure 6.13 General concept of distribution of winding's partial parasitic capacitances.

eralized findings. Most of the conducted emission frequency range is covered by capacitive character of the used inductors. Inductive character is commonly observed only below few hundreds of kilohertz, where the inductor's main self-resonance $f_{Rp,L}$ usually occurs. Exemplary impedance characteristic of the evaluated inductor within the frequency range of conducted emission is presented in Figure 6.12.

In the frequency range above the inductor's main self-resonance $f_{Rp,L}$, many of resonances are possible to occur, but usually only few of them are meaningful for the resultant impedance characteristic, and they are quite distant from each other in the frequency domain. Therefore, a circuit model development can be limited to only those mostly exposed resonances and founded on their frequencies $f_{Rp,L1}$, $f_{Rp,L2}$, ..., which can be graphically estimated based on visual overview of the impedance characteristic. From the point of view of winding structure, it can be expected that partial parasitic capacitances are distributed along winding, forming a specific inductance–capacitance chain presented in Figure 6.13. This presumption is most adequate for one-layer windings with the grounded core.

Ladder circuit models of windings are used for analysis of voltage distribution in high-power transformer windings during fast voltage transients caused by lightning. Furthermore, ladder circuit models are used for simplified analysis of transmission line effects, especially transmission lines with inhomogeneous parameters. Arrangements of winding's layers of commonly manufactured inductors are usually more complex than one-layer design, which results with more inhomogeneous distribution of partial parasitic capacitances as compared to one-layer windings structure.

An equivalent parasitic capacitance of the entire inductor can be estimated based on the impedance-frequency characteristic within frequency range in which capacitive character of the inductor is dominating–between the major parallel resonance $f_{Rp,L}$ and major serial resonance $f_{Rs,L}$, and where the phase of impedance is negative (Figure 6.6 [20,21]. Calculations carried out for the evaluated inductor (Figure 6.8) show that inductor's equivalent capacitance is not constant, slightly increases with the increase of frequency, which is most

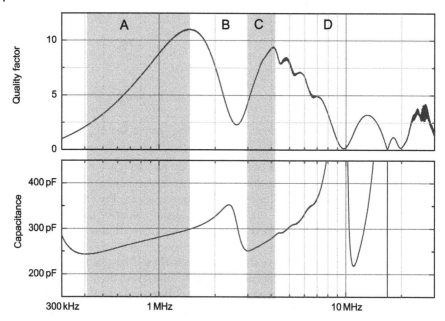

Figure 6.14 Changes of equivalent parasitic capacitance and associated quality factor versus frequency.

clearly visible between the major self-resonance $f_{Rp,L}$ and the minor resonance frequencies $f_{Rs,L1}$, $f_{Rs,L2}$ in Figure 6.14—the frequency subranges A and C.

This is an effect of the increase of dielectric losses in windings insulation material with the increase of frequency that is quantified as the dielectric loss factor $tg\delta$ (6.10), where ε and ε'' are real and imaginary components of dielectric permittivity. Especially, imaginary part of permittivity is frequency dependent, thus it changes the effective parasitic capacitance [22,23].

Inductor's dielectric losses increasing in the frequency range where capacitive character dominates can be modeled using a ladder circuit model constructed in a similar way as the one used for modeling the increase of losses in windings (Figure 6.9). This is because parallel configurations are preferable for easier identification of model parameters (Figure 6.15) [12,14].

$$tg\delta = \frac{1}{Q} = \frac{\varepsilon}{\varepsilon''} \tag{6.10}$$

More considerable changes of effective parasitic capacitance occur within frequency ranges close to the less expressive resonances $f_{Rs,L1}$, $f_{Rs,L2}$ marked as the subranges B and D in Figure 6.14. This is an effect of distributed character of partial parasitic capacitances associated with resonances in particular internal sections of a winding. Due to local resonances, in particular winding sections, for frequencies above these resonance frequencies, groups of partial capacitances associated with the resonances become less meaningful for the resultant capacitance, because inductive reactances of these sections become locally dominant.

Changes of the resultant equivalent parasitic capacitance with frequency, associated with the internal minor resonances and increasing HF losses, result in the fact that inductor's circuit model—in which all parasitic partial capacitances are represented by only one equivalent capacitance connected to external inductor's terminals—is not sufficient to obtain appropriate accuracy for many inductors. Therefore, to improve adequacy of the model, topology should be expanded by extra internal components, taking into account that the exploration of the inductor's internal structure is not accessible for identification tests.

Theoretically, each impedance—frequency characteristic can be modeled by fitted mathematical formulas, using number of methods, for example, vector fitting [18]. One-port network approach can be also implemented based on measurement results of external impedance characteristics only. Although these approaches allow for limited use of simulation tools for circuit modeling, they do not provide much information about internal physical behavior of the modeled component, as these models are of a "black box" type.

One of the possible solutions is to fit equivalent circuit topology, which will be beneficial for behavioral analysis, and to find parameters of such a circuit model based on analysis of external impedance characteristics only. Based on the comparative analysis of measured impedance characteristics of a number of inductors of different sizes and commonly known behavioral recognition concerning parasitic capacitances, the first Cauer topology has been proposed for modeling inductor's internal resonances [24] and the second Foster topology for modeling parasitic capacitance changes induced by dielectric losses that increase with the frequency [22,23].

According to the first Cauer topology, the inductor's equivalent parasitic capacitance can be replaced by adequately long ladder of inductances L_{R1}, L_{R2}, L_{R3}, ... and capacitances C_{R1}, C_{R2}, C_{R3}, ... presented in Figure 6.15. Impedance of the last ladder rung R'_{DC}, where L'_{AB} is a load, form the point of view of Cauer synthesis, but from the point of view of the evaluated inductor' model it is a part of a circuit model representing inductive part of the impedance characteristic, below the main self-resonance $f_{Rp,L}$.

Parameters R'_{DC}, L'_{AB} are not directly equal to those defined in Figure 6.15, R_{DC}, L_{AB}, because inductances L_{R1}, L_{R2}, L_{R3}, ... added to the model with associated resistances $R_{R,L1}$, $R_{R,L2}$, $R_{R,L3}$, ... should also be taken into account for calculating effective inductance and resistance at LF range according to formulas (6.11) and (6.12).

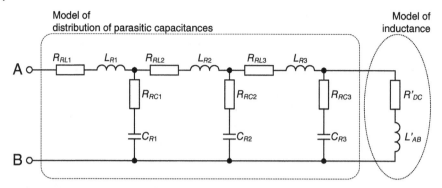

Figure 6.15 Circuit model of distributed parasitic capacitances of an inductor—the First Cauer topology.

$$L_{AB} \approx \sum_{n=1}^{n} L_{R,Ln} + L'_{AB} \qquad (6.11)$$

$$R_{DC} \approx \sum_{n=1}^{n} R_{R,Ln} + R'_{DC} \qquad (6.12)$$

Each rung of such a configured ladder circuit model is intended to represent one pair of coupled internal resonances, parallel $f_{Rp,L1}$, $f_{Rp,L2}$, $f_{Rp,L3}$, ... and serial $f_{Rs,L1}$, $f_{Rs,L2}$, $f_{Rs,L3}$, ... that can be identified at the impedance characteristic, for example, in Figure 6.12. On the other hand, it can be noticed that windings behavior in HF range, where capacitive character is dominating, can also be considered as specific transmission lines (TL) with unequally distributed inductances and capacitances. These inhomogeneities in distribution of parasitic capacitances associated with particular winding's structure components can be represented by separate ladder rungs of different parameters L_{R1}-C_{R1}, L_{R2}-C_{R2}, and so on (Figure 6.15).

According to the Cauer circuit synthesis, input impedance can be formulated analytically as continued fraction from (6.13), omitting losses represented by adequate resistances and the loading impedance $Z_L = R'_{DC} + j\omega L'_{AB}$.

$$Z(s) \approx sL_{R1} + \cfrac{1}{sC_{R1} + \cfrac{1}{sL_{R2} + \cfrac{1}{sC_{R2} + \dots + \cfrac{1}{sL_{RN} + sC_{RN}}}}} \qquad (6.13)$$

On the other hand, the same impedance can be expressed by the use of polynomial form with zeros and poles that correspond to critical frequencies of serial and parallel resonances (6.14).

$$Z_{AB}(s)\Big|_{s=j\omega} = jX(\omega) = jk\frac{\omega(\omega_{z1}^2 - \omega^2)(\omega_{z2}^2 - \omega^2)\dots(\omega_{zN}^2 - \omega^2)}{(\omega_{p1}^2 - \omega^2)(\omega_{p2}^2 - \omega^2)\dots(\omega_{pN}^2 - \omega^2)} \qquad (6.14)$$

Parameters of components of the Cauer equivalent circuit can be initially estimated based on frequency values of the resonances possible to detect from the experimentally obtained impedance characteristic [25,26]. Then to achieve a better adequacy of the impedance characteristic, initial parameters of the model can be optimized numerically using error minimizing algorithms [27].

This method successfully allows finding parameters of the circuit model, however, there are some inconveniences with the Cauer circuit model synthesis. First, the Cauer synthesis method is ambiguous thus many solutions are possible, and second, not all transfer functions are possible to solve using this method. Therefore, for successful effects it is essential to keep the order of the proposed circuit model as low as possible and to select starting point parameters of the model closely correlated to experimentally obtained characteristics.

Simplified graphical correlation between capacitive part of the inductor's impedance characteristic, the ladder circuit model parameters, and internal resonance frequencies is presented in Figure 6.16. Referring to this correlation, graphical analysis of the experimentally obtained impedance characteristics makes it possible to successfully estimate the required number of ladder circuit nodes and initial parameters of each of rungs that are well enough correlated to the most meaningful physical features of the modeled inductor. Individual sections of linear sections of impedance characteristic with the slopes +/-20 dB, which are usually easy to detect, directly suggest values of parameters of components of resonance circuits associated with particular ladder circuit rungs. This method is especially effective in a case where nearby resonance frequencies are relatively distant and thus more expressive. If resonance frequencies are close to each other, their visibility on the impedance characteristic, depending on quality factor of a resonance circuit, can be less clear and may often require more advanced methods for numerical optimization. An example of the ladder

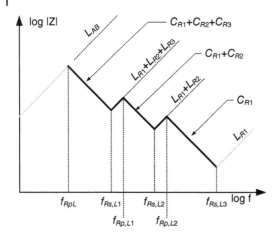

Figure 6.16 Simplified theoretical characteristic of input impedance of LC ladder circuit.

circuit model developed for the evaluated inductor of nominal inductance 10 mH and rated current 16 A is presented in Figure 6.17. In this model, the changes of inductor's inductance with frequency are modeled by three sectional Foster circuit comprising inductances connected in parallel with resistances (500 Ω and 2.2 mH, 150 Ω abd 2.5 mH, 50 Ω and 3.5 mH) and three sectional Cauer circuits topology comprising inductances and capacitances (450 nH and 220 pF, 7 μ H and 43 pF, 100 μ H and 35 pF) with losses represented by adequate resistances.

Comparison of the measured impedance characteristic and the calculated one using the developed circuit model is presented in Figure 6.18. The achieved results can be accepted as satisfactory up to the frequency of the main serial resonance between 10 and 20 MHz–as it was intended choosing such simplified topology with only three rungs. To achieve better adequacy in the frequency range above this resonance, subsequent ladder sections of the inductance model should be added.

The model accuracy can also be improved by more accurate representation of the increase of the equivalent parasitic capacitance and dielectric losses with frequency (Figure 6.14) using an additional ladder circuit consisting of capacitances and resistances in the arrangement shown in Figure 6.19.

Nevertheless, as it was underlined in previous chapters, in many cases keeping model simplicity within reasonable limits can be more profitable than small increase of accuracy. Therefore, every additional model extension should be carefully considered, taking into account expected overall advantages and disadvantages. Even less accurate but physically more adequate models can be more beneficial for further analysis than more accurate models but without physical understanding, because the former allow for more practical further

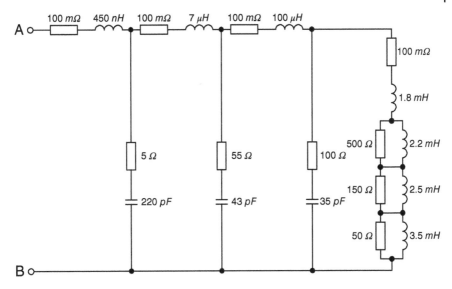

Figure 6.17 Topology and parameters of the developed ladder circuit model of inductor's DM impedance taking into account variable inductance and the most expressive internal parasitic resonances.

conclusions. For example, based on the developed circuit model, which is physically understandable, inductive and capacitive current components can be determined (Figure 6.20).

Decomposition of the inductor's current into capacitive and inductive parts demonstrate that below the frequency of the main parallel self-resonance (which is about 200 kHz), the capacitive current component increases with frequency with the slope of about 20 dB/dec (line a), whereas the inductive current component decreases much faster above this resonance frequency. Decrease of inductive component of the inductor's current is about 40 dB per decade (line b) up to the first internal resonance, about 80 dB per decade (line c) between the first and the second internal resonance, and even more in frequency ranges above the further internal resonances. Presented exemplary analysis based on the developed circuit model representation allows concluding that internal resonances of inductor's windings resulting from distributed character of parasitic capacitances significantly reduce influence of inductor's equivalent inductance on the impedance characteristic above frequency of the main parallel parasitic self-resonance.

6.3.3 Circuit Modeling of Inductor's CM Capacitances

The circuit model presented in the previous chapter was focused on impedance characteristics of inductor used as a DM component (Figure 6.1a). The analysis

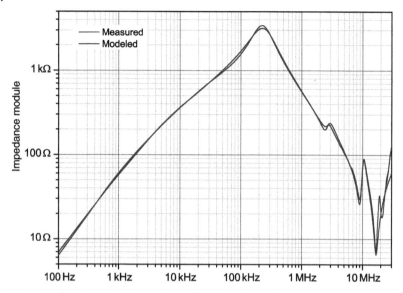

Figure 6.18 Modeled and experimental DM impedance–frequency characteristics of the evaluated inductor.

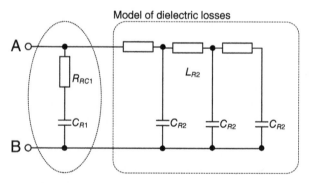

Figure 6.19 Ladder circuit model of dielectric losses increasing with frequency associated to parasitic capacitances.

of propagation of CM currents requires to extend this model by additional capacitances representing inductor-to-ground parasitic capacitive interactions (Figure 6.1b).

Detailed categorization of all distributed parasitic partial capacitances into internal windings capacitances and winding-to-ground capacitances is difficult and requires to use laborious and invasive field-oriented methods. Such an approach is usually not reasonable, taking into account identification efforts

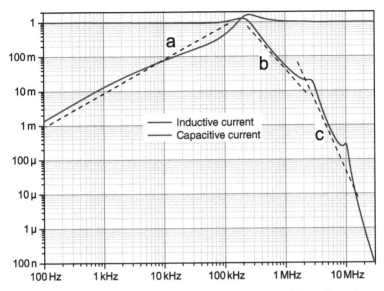

Figure 6.20 Inductive and capacitive current components of the evaluated inductor–relative comparison.

and limited increase of accuracy cased by further simplifications that are usually necessary.

Simplified approach to extraction of inductor's winding-to-ground capacitances from all parasitic capacitances can be based on a circuit model limited only to inductor's external terminals (Figure 6.5). With such an assumption, only two winding-to-ground parasitic capacitances associated with the inductor terminals A and B are possible to define, $C_{A,Gnd}$ and $C_{B,Gnd}$ (Figure 6.21). Consequently, the total parasitic capacitance of inductor in relation to ground can be easily measured according to the test setup presented in Figure 6.21a, where both inductor's terminals are shorted, thus the measured capacitance in this test configuration is a sum of $C_{A,Gnd}$ and $C_{B,Gnd}$ (6.15).

$$C\Big|_{Z_{AB-Gnd}} = C_{A,Gnd} + C_{B,Gnd} \tag{6.15}$$

Adequacy of such a simplified circuit model for analysis of CM currents in frequency range of conducted emission can be verified by analysis of inductor's impedance characteristics measured in test setups with different grounding configurations. The inductor's CM impedance characteristic Z_{AB-Gnd}, measured using test setups presented in Figure 6.21, is presented in Figure 6.22. According to this impedance characteristic presented together with the inductor's DM impedance Z_{AB}, the winding-to-ground parasitic capacitance is

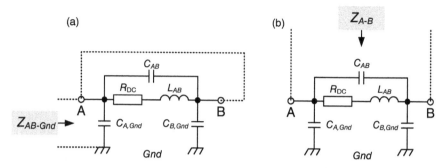

Figure 6.21 Test setups for measurement of inductor's CM and DM impedances.

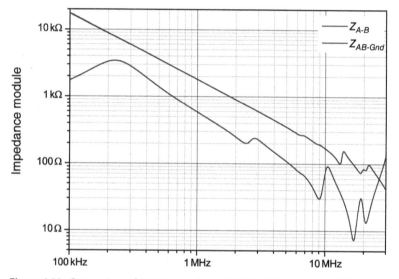

Figure 6.22 Comparison of CM (Z_{AB-Gnd}) and DM (Z_{AB}) impedances of the evaluated inductor.

significantly lower than the winding self-capacitance and less influenced by internal resonances within the frequency range up to the main serial resonance, which in the evaluated case is above 10 MHz.

The inductor's winding-to-ground parasitic capacitance is only slightly influenced by resonances located above the main serial parasitic resonance, which means that the circuit model of ground capacitances limited to external terminals can only be adequate in a wider frequency range than for inductors self-capacitance.

The overall winding-to-ground parasitic capacitance $C_{Z_{AB-Gnd}}$ measured using the test configuration presented in Figure 6.21a is usually not equally dis-

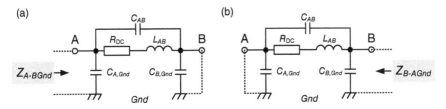

Figure 6.23 Test setups for measurement of unbalance of inductor's CM capacitances.

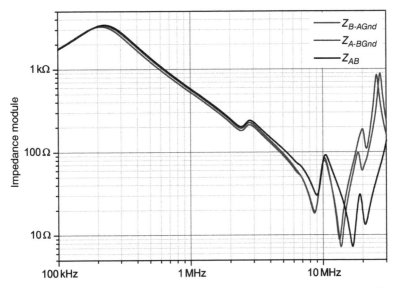

Figure 6.24 Inductor's impedance characteristics seen from both winding ends A and B used for estimation of unbalance of inductor's ground capacitances.

tributed between the inductor's terminals A and B. This is a result of commonly used winding structure, where both of winding ends are located in the opposite sides in relation to grounded structures, especially the inductor's core. Unbalanced distribution of winding-to-ground parasitic capacitances assigned to particular inductor's terminals can be estimated by measuring equivalent capacitances seen from both winding ends, with specific grounding connections made according to test setups presented in Figure 6.23. Inductor's impedance characteristics obtained in those test setup configurations, where one of the inductor's terminals is connected to ground, are shown in Figure 6.24 as Z_{A-BGnd} and Z_{B-AGnd} with comparison to the DM impedance Z_{AB}.

Shunting inductor's terminals to ground during the test changes winding-to-ground equivalent capacitance, thus the equivalent parasitic capacitances determined using different test setups allow defining adequate dependences between modeled capacitances Z_{A-BGnd}, Z_{B-AGnd}, and Z_{AB}. According to the circuit model (Figure 6.23), the equivalent capacitances determined in both configurations Z_{A-BGnd} and Z_{A-BGnd} are the sums of adequate capacitances of the circuit model (6.16) and (6.17).

$$C\Big|_{Z_{A-BGnd}} = C_{AB} + C_{A,Gnd} \tag{6.16}$$

$$C\Big|_{Z_{B-AGnd}} = C_{AB} + C_{B,Gnd} \tag{6.17}$$

The capacitance determined using DM setup configuration (Figure 6.23 b) can be formulated as (6.18) and is dependent on all capacitances included in the model C_{AB}, $C_{A,Gnd}$, and $C_{B,Gnd}$.

$$C\Big|_{Z_{A-B}} = C_{AB} + \frac{C_{A,Gnd}C_{B,Gnd}}{C_{A,Gnd} + C_{B,Gnd}} \tag{6.18}$$

The impedance characteristics presented in Figure 6.24 show that differences of inductor's impedances measured using different test setups become significant particularly around and above the main serial parasitic resonance frequency of the inductor, which in the evaluated case is above 10 MHz. It can also be noticed that shorting of one of the inductor's terminals to ground decreases the frequency of the main serial resonance, which means that the resultant equivalent parasitic capacitance of the inductor measured in these configurations increases in relation to capacitance measured in DM configuration (Figure 6.21b).

In the frequency range below 10 MHz, these differences are much smaller and more regular, but enough to determine unequal distribution of equivalent ground capacitances between terminals $C_{A,Gnd}$ and $C_{B,Gnd}$ that differentiate input and output impedances of the inductor represented as a two-port network (Figure 6.21b). Therefore, the inductor as CM current filter component is also slightly directional. The significance of the inductor's attenuation directivity can be particularly important in frequency ranges where the source and load impedances of the filtered system seen by filter are relatively high, thus should be more carefully correlated to unbalance of the inductor's terminals.

Finally, two-port circuit model of the inductor adequate for analysis of CM current propagation along with extracted parameters is presented in Figure 6.25. Having such a model, physically adequate and verified experimentally, propagation of CM conducted emission can be analyzed in different systems, especially where input and load impedances vary significantly, for ex-

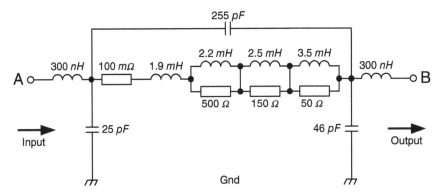

Figure 6.25 An example of two-port circuit model of the developed inductor with extracted parameters, relevant for analysis of CM currents propagation.

ample, interactions between the inductor as an RFI filter and the motor feeding cable.

6.4 Broadband Performance of AC Motor Windings

Windings of an AC motor as an inductive part have much more complex structure as compared to a simple inductor with single winding discussed in the previous chapter [28–31]. First, a set of AC motor windings consists of three similarly composed windings preliminary assigned to be connected to three phases of power system and can be configured into star or delta topology. Second, three-phase windings are spread around a cylindrical stator according to various distribution patterns dependent on the winding type, but in each type phase windings overlap each other as it is presented in a simplified form in Figure 6.26. Third, motor windings are wound differently than inductor's windings and therefore do not have so clearly layered structure that results in lower homogeneity between winding segments. Fourth, magnetic couplings between winding sections are slightly variable because of interactions resulting from rotation of wound rotor with different speeds. Therefore, detailed modeling of AC motor stator windings within wide-frequency range is much more complex than that of a simple inductor and requires more reasonable simplifications [32–34].

Similarly, as it was argued in previous sections for an inductor modeling, finite element method approach that allows taking all parasitic couplings into consideration, usually leads to the need of determination of enormous number of parameters. These parameters are tightly correlated to the internal structure of AC motor's windings that are usually not accessible for identification at the motor installation stage. Thus, terminal referenced simplified circuit models

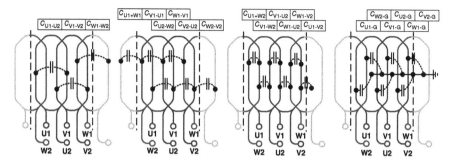

Figure 6.26 Simplified flat view of stator windings diagram of three phase AC motor with different categories of parasitic capacitive couplings between windings indicated.

with limited number of parameters are highly preferable for analysis of EMC behavior of an AC motor as one of the key components of an ASD. Taking into consideration the frequency range of conducted emission, commonly used topology of motor windings have parasitic capacitances of motor windings to ground essential for CM currents generation at the output side of ASD [35,36]. Therefore, development of a rationally simplified broadband circuit model of motor's windings along with an appropriate method of identification of its parameters based on results of impedance measurement via only externally accessible windings' terminals are very advisable.

6.4.1 Lumped Representation of Parasitic Capacitances of AC Motor Windings

An equivalent representation of distributed parasitic capacitances of motor windings can be initially limited to only six windings' terminals and the reference ground represented by the motor stator. Such a general approach allows defining 21 equivalent lumped parasitic capacitances presented in Figure 6.26 that can be categorized according to their physical meaning as self, interwinding, and ground capacitances (Table 6.1).

Referring to the analysis presented for a simple structured inductor it should be expected that parasitic capacitances of a single winding of the motor cannot be symmetrical in relation to their terminals—beginnings $V1, U1, W1$ and ends $U2, V2, W2$. This is an effect of slightly different allocations of particular winding's terminals in relation to the motor's stator and two other windings resulting from a specific structure of three phase windings distribution around the stator (Figure 6.26). Therefore, ground capacitances and interwinding capacitances are classified into subcategories, depending on the winding's terminals, beginnings, or ends. Circuit model of such defined AC motor windings with parasitic capacitances, referenced only to the winding terminals is presented in

Table 6.1 Categories of lumped parasitic capacitances of AC motor windings

Capacitance type	Symbol	Representatives
Self	C_{Self}	$C_{U1-U2}, C_{V1-V2}, C_{W1-W2},$
Inter-winding Type 1	C_{Int1}	$C_{U2-V1}, C_{V2-W1}, C_{W2-U1}$
Inter-winding Type 2	C_{Int2}	$C_{U1-V1}, C_{V1-W1}, C_{W1-U1}$
Inter-winding Type 3	C_{Int2}	$C_{U2-V2}, C_{V2-W2}, C_{W2-U2}$
Inter-winding Type 4	C_{Int2}	$C_{U1-V2}, C_{V1-W2}, C_{W1-U2}$
Ground Type 1	C_{Gnd1}	$C_{U1-Gnd}, C_{V1-Gnd}, C_{W1-Gnd},$
Ground Type 2	C_{Gnd2}	$C_{U2-Gnd}, C_{V2-Gnd}, C_{W2-Gnd},$

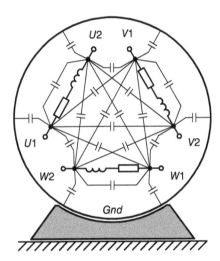

Figure 6.27 Lumped representation of parasitic capacitances of AC motor windings referenced to the motor terminals.

Figure 6.27, where all categories of parasitic capacitances are connected to the motor terminals.

The presented circuit model can be understood as a general six-port terminal model, which allows defining a matrix of capacitances describing all interterminals mutual capacitive couplings (Table 6.2). Unfortunately, identification of all capacitances of such six-port model is rarely possible because distributed character of represented capacitive couplings is usually too significant to be neglected.

Such a generic model of the AC motor can be useful in a wide range of applications, nevertheless identification of its parameters is difficult, because it includes 21 mutually related equivalent parasitic capacitances that cannot represent ad-

Table 6.2 Matrix of lumped parasitic capacitances of AC motor windings referenced to motor terminals

Motor terminal	U1	U2	V1	V2	W1	W2	Gnd
U1	-	C_{U1-U2}	C_{U1-V1}	C_{U1-V2}	C_{U1-W1}	C_{U1-W2}	C_{U1-Gnd}
U2	-	-	C_{U2-V1}	C_{U2-V2}	C_{U2-W1}	C_{U2-W2}	C_{U2-Gnd}
V1			-	C_{V1-V2}	C_{V1-W1}	C_{V1-W2}	C_{V1-Gnd}
V2				-	C_{V2-W1}	C_{V2-W2}	C_{V2-Gnd}
W1					-	C_{W1-W2}	C_{W1-Gnd}
W2						-	C_{W1-Gnd}
Gnd						-	-

equately distributed partial parasitic capacitances of windings. Difficulty in a general approach of modeling of AC motor windings can be significantly reduced by taking into consideration that in practical use the motor windings in most of applications are connected into star or delta configuration, which limits the number of external motor's terminals from six to three only. Further analysis will be continued for star connection that is more commonly used and it is also easier to analyze within a broad band of frequency because only one of each winding's ends, $V1, U1, W1$, is connected to the externally accessible motor's terminals.

When searching for further reasonable simplifications of possible circuit representation of motor windings it can be noticed that from the point of view of analysis of conducted emission, only CM capacitances are supposed to be meaningful. Nevertheless, as it was explained in previous chapters, DM capacitances are also tightly associated with CM current generation paths because of the voltage and impedance unbalances commonly occurring at the output side of FC, which initiate DM-to-CM and CM-to-DM currents conversion phenomena. Preliminary analysis of AC motor impedance characteristics, measured via external motor terminals, can be helpful for advising further simplifications of the windings circuit model that allow finding reasonable compromise between the model adequacy and difficulty of identification of its parameters.

6.4.2 Impedance–Frequency Characteristics of AC Motor Windings

CM and DM impedance characteristics of motor with windings, connected into star or delta configuration seen from the external terminals can be relatively easily measured in two general test configurations presented in Figure 6.28. Adequate DM and CM impedance characteristics measured for AC motors of different rated power, from few kilowatts up to half megawatt, are presented in

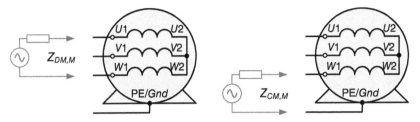

Figure 6.28 Test setup for measurement of DM and CM impedance characteristic of AC motor windings.

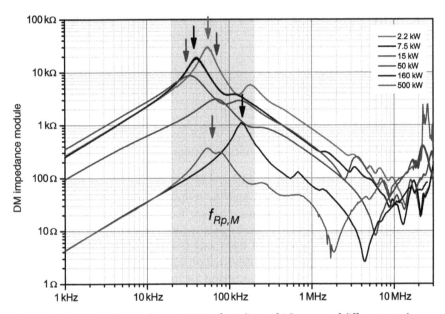

Figure 6.29 DM impedance characteristics of windings of AC motors of different rated power.

Figures 6.29 and 6.30. In all tested windings of AC motors, two dominating main resonances can be easily detected based on rough overview of the measured impedance characteristics: parallel resonances $f_{Rp,M}$ and serial resonances $f_{Rs,M}$.

Parallel main resonances $f_{Rp,M}$ are the easiest to estimate based on DM impedance characteristics as a frequency above which the dominating inductive character of windings changes into capacitive (Figure 6.29). For many tested motors of rated power up to megawatts, the frequencies of main parallel resonances are in most cases located in frequency range from few tens of kilohertz up to few hundreds of kilohertz. Not unambiguous dependence has been found

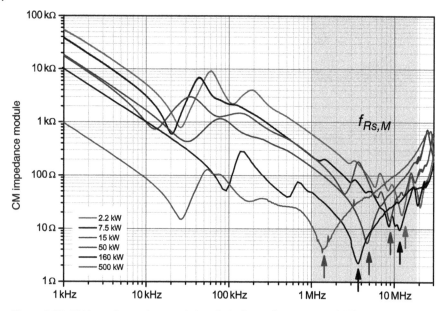

Figure 6.30 CM impedance characteristics of windings of AC motors of different rated power.

between the motor-rated power and the frequency of the main parallel resonance, which indicates that values of these frequencies are more dependent on windings' structures than on the motor-rated power; however, particular structures of windings have not been analyzed separately. Effects of parallel resonances of windings are also reflected on CM impedance characteristics as significant impedance changes occurring for frequencies closely correlated to $f_{Rp,M}$.

The main serial resonances of the motor's windings $f_{Rs,M}$ are most clearly visible at CM impedance characteristics as the frequencies for which the CM impedance becomes minimal, and above which capacitive character of CM impedance changes its character into inductive (Figure 6.30). For the tested range of rated power of AC motors, the detected frequencies of serial resonances were located in the frequency range above 1 MHz up to over a ten of MHz. It can be noticed that frequencies of winding's serial resonances are more regularly associated with the motors, rated power, and generally decrease with the size of motor, nevertheless in some cases of motors of similar power this rule may not be fulfilled.

In the frequency range between the main parallel $f_{Rp,M}$ and serial $f_{Rs,M}$ resonances, most of the winding's impedance characteristics measured for motors of different rated power expose a number of noticeable deformations that are a result of less meaningful serial and parallel resonances occurring in windings

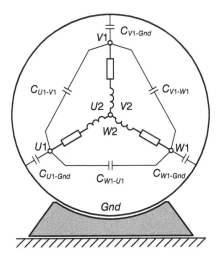

Figure 6.31 Reduced representation of lumped parasitic capacitances of AC motor windings connected into star topology.

locally, between internal subsections of windings. However, in frequency range above $f_{Rs,M}$, the observed impedance characteristics usually become more irregular, with much higher number of resonances that are also sharper than below $f_{Rs,M}$.

Less meaningful resonances are clearly visible, at DM and as well at CM impedances characteristics. Impedance irregularities confirm that internal parasitic resonances occurring inside windings are significant, and that circuit modeling limited to nodes associated with the external terminals of windings only will not be sufficient to adequately reflect these resonances.

6.4.3 Identification of Lumped Parasitic Capacitances of AC Motor Windings

Identification of parasitic capacitances and further analysis will be continued for one of the tested AC motors of rated power 7.5 kW. Windings of the tested motor were configured in star connection, thus number of the parasitic capacitances to be determined is significantly reduced in relation to the previously discussed general approach for separate windings. Terminal referenced circuit model of AC motor, reduced due to connection of winding into a star topology, contains only six equivalent capacitances: three DM C_{U1-V1}, C_{V1-W1}, C_{W1-U1} and three CM C_{U1-Gnd}, C_{V1-Gnd}, C_{W1-Gnd} connected to the motor's terminals according to Figure 6.31.

AC motor windings connected into a star topology expose better impedance symmetry between winding terminals $U1$, $V1$, and $W1$, because in the star

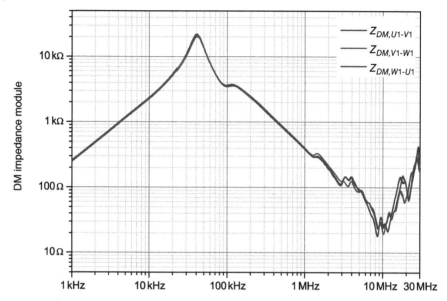

Figure 6.32 DM impedance characteristics measured for different phases of AC motor windings.

connection always only the same categories of windings terminals, beginnings or ends, are used as the external terminals of star topology. Symmetry between impedances of the motor's phase winding can be verified by a measurement of impedances in all possible DM and CM configurations; $Z_{DM,U1-V1}$, $Z_{DM,V1-W1}$, $Z_{DM,W1-U1}$ and $Z_{CM,U1}$, $Z_{CM,V1}$, $Z_{CM,W1}$. Adequate impedance characteristics measured for evaluated motor are presented in Figures 6.32 and 6.33, and show that only slight impedance imbalance can be noticed in the frequency range close to the main serial resonance $f_{Rs,M}$, which is about 10 MHz and above this frequency.

The balance between phase impedances of the motor significantly allows reduceing the identification efforts of parasitic capacitances. Nevertheless, the presented CM impedance characteristic shows that a simplified approach cannot be used in the entire frequency range of conducted emission because CM impedances change significantly in the frequency range correlated to the main parallel resonance $f_{Rp,M}$. Thus, the adequate equivalent parasitic capacitance is not constant but can change with frequency. Therefore, the discussed model can be satisfactorily adequate only in the frequency range between the main parallel resonance $f_{M,p}$ and the main serial resonance $f_{M,s}$, where the estimated equivalent capacitance can be recognized as approximately constant. It should also be noticed that in this frequency range noticeable changes of CM and DM impedance characteristics with frequency are very similar in shape,

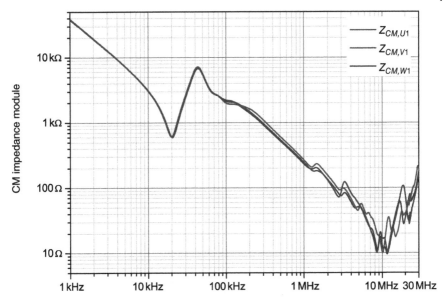

Figure 6.33 CM impedance characteristics measured for different phases of AC motor windings.

despite the fact that both characteristics are shifted by approximately a constant distanne in the logarithmic impedance scale. Such a correlation of CM and DM impedance characteristics confirm additionally the adequacy of the circuit model with constant CM and DM capacitances within frequency range between the main resonances, parallel $f_{M,p}$ and serial $f_{M,s}$.

Accepting such bandwidth limitations of the model, CM and DM equivalent capacitances can be identified based on measured differences between the CM and DM impedances and dependencies resulting from the circuit model presented in Figure 6.31. Relatively regular difference between CM and DM impedances within the frequency range between $f_{M,p}$ and $f_{M,s}$ is indicated in Figure 6.34 by dashed lines a and b.

According to Figure 6.31, formulas for parasitic capacitances determined using the DM test setup configuration can be formulated as (6.19) for phases U1 and V1 and similarly using the CM test setup as (6.20) for phase U1

$$C\Big|_{Z_{DM,U1-V1}} = C_{DM} + \frac{1}{2}\left(C_{CM} + C_{DM}\right) \tag{6.19}$$

$$C\Big|_{Z_{CM,U1}} = C_{CM} + \frac{1}{2}\left(\frac{C_{CM}C_{DM}}{C_{CM} + C_{DM}}\right) \tag{6.20}$$

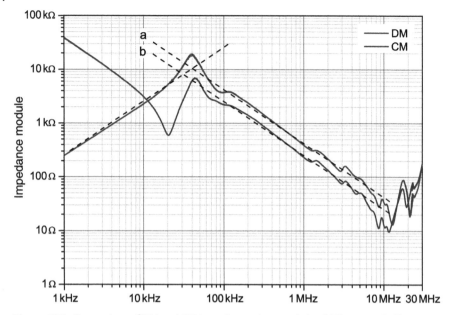

Figure 6.34 Comparison of DM and CM impedance characteristic of AC motor windings.

where $C_{DM} = C_{U1-V1} = C_{V1-W1} = C_{W1-U1}$ and $C_{CM} = C_{U1-Gnd} = C_{V1-Gnd} = C_{W1-Gnd}$.

The estimated values of equivalent parasitic capacitances for the tested AC motor of the rated power 7.5 kW, corresponding to the simplified circuit model presented in Figure 6.31, are equal, $C_{DM} \approx 96$ pF and $C_{CM} \approx 453$ pF.

6.4.4 Modeling of Distributed Parasitic Capacitances of AC Motor Windings

The impedance peaks and deeps visible on CM impedance characteristics of various AC motor windings presented in Figure 6.30 indicate that equivalent parasitic CM capacitances of windings seen via external terminals can be different depending on the measurement frequency. For all tested motors, numerous clearly noticeable deviations of impedance characteristic in relation to ideal impedance characteristic of pure capacitance, which is sloping -20 dB per decade, are present in the frequency range below the frequency of the main serial resonance $f_{Rs,M}$. Also in all tested motor's windings, the most pronounced deviations of impedance characteristics are tightly correlated to the frequency of the main parallel self-resonance $f_{Rp,M}$ of windings, which is the lowest frequency of occurring deviations. Step change of equivalent parasitic CM capacitance of the winding, associated with this deviation is presented in

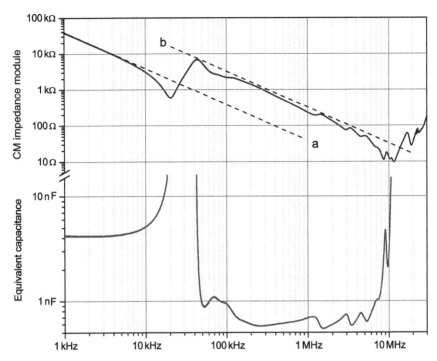

Figure 6.35 Changes of equivalent parasitic capacitances of AC motor windings with frequency.

Figure 6.35 as a distance between tangential, -20 dB sloping, lines a and b, which help to determine values of these capacitances.

The most pronounced change of the equivalent parasitic capacitance is an effect of main self-resonances of motor's windings due to which for higher frequencies some parts of windings are successfully separated from the rest of the winding structure by internally created resonance subcircuit, which acts as a specific low-pass filter. Therefore, parasitic capacitances of these parts become less meaningful for the resultant equivalent capacitance seen externally via winding terminals. The others, less expressed deformation of impedance characteristics are correlated to subsequent internal resonances of windings resulting from the winding particular structure.

Similarly, as it was explained for an inductor in Section 6.3, such a frequency-dependent impedance resulting from distributed character of partial parasitic capacitances can be successfully modeled using an LC ladder circuit with HF losses represented by appropriate resistances. General concept of distribution of partial parasitic capacitances of motor's windings connected into a star configuration can be represented as a ladder-related circuit, which takes into account CM and DM capacitances (Figure 6.36).

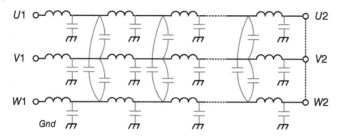

Figure 6.36 Distributed representation of CM and DM parasitic capacitances of AC motor windings configured in star connection.

On the basis of analysis of experimentally obtained DM and CM impedance characteristics of motor's windings, ladder circuit models for analysis of DM and CM phenomena can be individually developed, similarly as for an inductor (Section 6.3). Nevertheless, simultaneous identification of both types, CM and DM, of distributed parasitic capacitances is much more difficult, because the developed behavioral individual models for CM and DM are highly simplified, and may not fully accurately reflect all the actual behaviors, especially if equivalent parasitic capacitances identified individually for DM and CM models are directly integrated into one aggregated model. Therefore, according to considerations related to commonly used switching patterns in FCs presented in Chapter 2, to achieve a better adequacy of ladder circuit models, it is recommended to analyze DM and CM phenomena separately.

A separate approach to DM and CM analysis is naturally simplified in ASDs, because frequency bands of domination of DM and CM phenomena being the main subject of interests, are relatively different. DM operating currents are of much lower frequencies than the CM currents, which are analyzed up to the frequency of 30 MHz. Additionally, generated CM transients are usually much shorter in relation to time constant of changes of operating DM currents, which therefore can be recognized as constant during the entire single process of generation of CM currents .

As demonstrated in Chapter 3, CM current generation is mainly associated with the output voltage switching at only one of converter's terminals at a time, thus CM impedance of motor windings, determined between one of winding's terminals in relation to ground (Figure 6.28a), is mostly appropriate for analysis of generation of conducted emission at the output side of FC. Therefore, further development of a winding circuit model will be continued considering primarily CM phenomena.

Considering that single generation process of CM currents is associated with voltage switching at one terminal at a time only, a representation of all parasitic capacitances of winding (Figure 6.36), CM and DM, can be simplified to one-port configuration presented in Figure 6.37. Unequally distributed partial

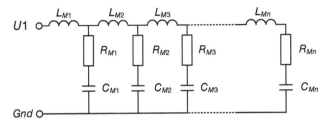

Figure 6.37 One-port representation of distributed parasitic capacitances of AC motor windings.

parasitic capacitances between motors windings and ground are represented by a number of steps consisting of lumped LC components with losses expressed by resistances.

This model can be misleadingly understood as representing CM distributed capacitances of windings of only one-phase motor winding. It should be underlined that in such an approach all distributed parasitic capacitances of all motor windings are taken into account, although not directly.

The total CM current $I_{CM,V1}$ generated at the energized motor's terminal is a sum of CM currents directly associated with distributed ground capacitances of energized phase winding $I'_{CM,V1}$ and another component $I''_{CM,V1}$ that is a CM current flowing toward ground via DM capacitances between windings of different phases and ground capacitances of two other phases (6.21).

$$I_{CM,V1} = I'_{CM,V1} + I''_{CM,V1} \qquad (6.21)$$

Graphical representation of summation of these two components, $I'_{CM,V1}$ and $I''_{CM,V1}$ of the total CM current is visualized in Figure 6.38. According to the perception of CM currents seen via one motor terminal presented in Figure 6.38, CM capacitances determined based on the CM impedance characteristics for one phase of motor's windings are also influenced by parasitic capacitances of the other two windings. This is also one of the explanations why three CM models obtained for different phases of motor windings cannot be directly integrated into one three-phase model of the AC motor. Despite such complex dependencies between CM and DM phenomena, modeling of CM currents based on CM impedance determined for one motor's terminal most adequately represent actual CM current generation process during single commutation of the FC output voltage.

The theoretical, simplified impedance characteristic of the circuit model of motor windings (Figure 6.37) is presented in Figure 6.39. Series of pairs of parallel and serial resonance frequencies occurring at the winding's CM impedance characteristic can be correlated with the number of steps in the ladder circuit

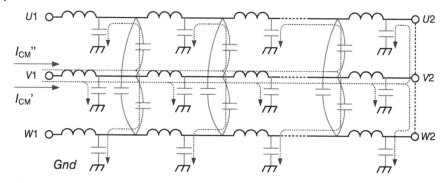

Figure 6.38 CM currents distribution in an LC ladder circuit model of AC motor windings.

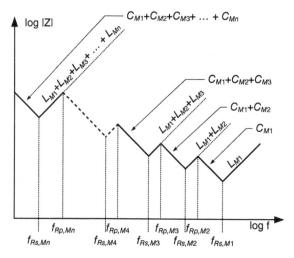

Figure 6.39 Theoretical CM impedance characteristic of AC motor windings represented by the LC ladder circuit model.

model and subsequently to LC parameters of these steps, according to formulas assigned to linear sections of the impedance characteristic.

Having the experimentally obtained CM impedance characteristic of the motor and the proposed ladder circuit model, initial values of parameters of LC components can be estimated based on detected perceivable resonance frequencies, similarly as it was presented for an inductor in Section 6.3.2. Afterward these parameters can be optimized numerically using error minimizing algorithms, which also allow to find appropriate resistances that represent HF losses of each ladder rung. Parameters of the ladder circuit model, obtained from 7.5 kW AC motor using this method, based on one of CM impedance characteristics (Figure 6.33), are presented in Figure 6.40.

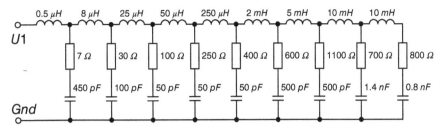

Figure 6.40 Parameters of the ladder circuit model determined for 7.5 kW AC motor.

The proposed nine-rung model corresponds directly to nine serial resonances, $f_{Rs,M1} \cdots f_{Rs,Mn}$, which represent the local decline of the impedance at the motor winding. Furthermore, higher quantity of parallel resonances in comparison to the number of ladder rungs can exist in the evaluated circuit model. Nevertheless, usually only parallel resonances occurring in nearby rungs are mainly meaningful and visible in the overall impedance characteristic $f_{Rp,M1} \cdots f_{Rp,Mn}$, which are tightly related to each other, for example, $f_{Rs,M2}$ and $f_{Rp,M2}$. It should be underlined that resonance frequencies can be very weakly visible at some parts of experimentally obtained impedance characteristics when quality factor of the correlated resonance circuits is adequately low. Such a case can be observed at the evaluated impedance characteristic (Figure 6.35) in the frequency range of about 100 kHz, where only the slope of the impedance characteristic, higher than -20 dB/dec, suggests existence of extra resonances that should be represented by extra rungs in the ladder circuit model.

Comparison of the modeled and experimentally obtained CM impedance characteristics is presented in Figure 6.41. This comparison clearly shows the correspondence between resonances and deformation of the resultant impedance characteristic. Existing discrepancies between the modeled and experimental characteristics can be further minimized; however, those presented—not fully optimized example—more clearly explains correlations between internal resonances, especially those that are less visible and less meaningful for resultant changes of the impedance characteristic.

Generally, taking into account that in real ASD application many factors influencing CM current generation that are difficult to predict may exist, the proposed broadband model of motor windings may be considered as sufficiently adequate for efficient estimation of generated CM currents. The main advantage of the proposed approach is a relatively easy procedure of model identification that is effective for AC motors of different rated power and require impedance measurements only via externally accessible motor's terminals [37].

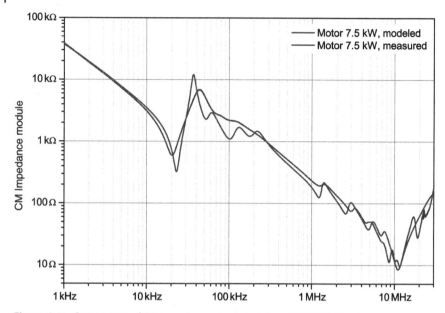

Figure 6.41 Comparison of CM impedance characteristics of 7.5 kW AC motor windings–measured and calculated using the developed ladder circuit model.

6.5 Broadband Performance of Motor Feeding Cable

The motor feeding cable in an ASD is a power cable with at least three active wires that are used for connecting the FC output terminals with the AC motor terminals. This specific kind of power connection is extra demanding from the point of view of an ASD operation and power transfer as well, because it is continuously energized by high-speed voltage switchings generating very wide spectrum of voltage harmonics reaching a bandwidth of several megahertz [33,38]. Transferring of such HF components voltage of relatively high magnitudes requires taking parasitic parameters of power cables into consideration, which are ignored in most of power applications. Therefore, broadband behavior of power cables used for feeding AC motors in ASDs must be considered in analysis of EMC performance of ASDs as one of the key components responsible for the resultant EM emission.

6.5.1 Broadband Behavior of Single Wires

Each electric wire manifests distributed parasitic capacitances in relation to ground and self-inductances that influence its impedance characteristic in HF range, and finally results in occurrence of transmission line (TL) effects at adequately high frequencies. The broadband characteristic of a single wire

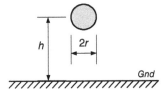

Figure 6.42 Geometry of transmission line formed by a single wire over the ground.

used for transfer of signals referenced to ground depend on values of parasitic capacitances and inductances that are substantially dependent on the distance between the wire and the ground plane. Per-unit parameters of a single wire of radius r located over the ground plane at height h (Figure 6.42) can be determined using formulas given in Ref. 2: inductance L'_C (6.22) and capacitance C'_C (6.22).

$$L'_C = \frac{\mu}{2\pi}\cosh^{-1}\left(\frac{h}{r}\right) \quad (H/m) \tag{6.22}$$

$$C'_C = \frac{2\pi\varepsilon_0}{\cosh^{-1}\left(\frac{h}{r}\right)} \quad (F/m) \tag{6.23}$$

According to these formulas and assuming that the wire diameter is usually foremost matched to the carrying current, the broadband characteristic of a single wire is mostly dependent on distance h between the wire and the ground, omitting relative permittivity of the used isolation. Simplified relation, for lossless case, between per-unit parameters of a single wire L'_C and C'_C along with the resultant characteristic impedance Z_0, calculated according to formula (6.22) is presented in Figure 6.43 as a function of relative height h/r.

$$Z_{0C} = \sqrt{\frac{L'_C}{C'_C}} = \frac{1}{2\pi}\sqrt{\frac{\mu}{\varepsilon_0}}\cosh^{-1}\left(\frac{h}{r}\right) \tag{6.24}$$

Presented analysis shows that for commonly used arrangements of motor cables in ASD, where mostly encountered distances between the cable wires and the grounded fixture are usually from fractions to several tens of wire radius, expected values of the characteristic impedance can be up to more than hundreds of ohms. Furthermore, the wire characteristic impedance is logarithmically dependent on the wire distance to the reference ground, which can result in significant inhomogeneities of the characteristic impedance due to arrangement of unshielded motor cables. Shielded cables are less sensitive

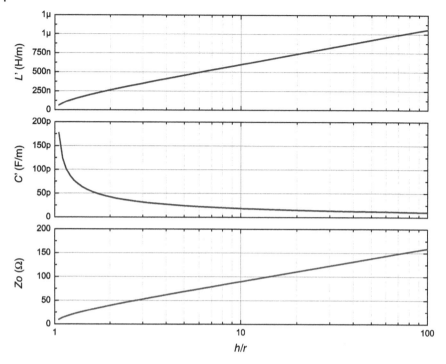

Figure 6.43 Per-unit parameters of single wire over grounded conducting plane: inductance L'_C, capacitance C'_C, and characteristic impedance Z_{0C}.

for arrangement in relation to the ground, as the shield of motor feeding cable should be efficiently grounded, which effectively equalizes the distances between cable wires and the ground. Therefore, the analysis of broadband behavior of motor cables will be continued only for shielded cables, as they are increasingly used in ASDs because of EMC requirements and because ensuring the uniform placement of unshielded long motor cable in relation to the ground during testing is difficult [39,40].

6.5.2 Electrically Long or Short Wires

Referring to the previous section, electric wires evenly spaced over the ground can be represented as series of equally distributed inductances L'_C and capacitances C'_C, representing per-unit parameters, arranged in a ladder-like circuit, similarly as it was proposed for inductors and motor windings (Figure 6.44).

Equivalent total capacitance of the entire wire is a sum of all partial capacitances that can be calculated as $C_C = l_C C'_C$, where l_C is the length of the wire in meters. The cable capacitance C_C can also be easily determined experimentally

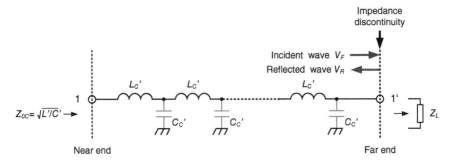

Figure 6.44 Ladder representation of equally distributed parasitic capacitances and inductances of a wire.

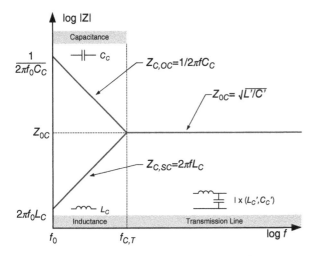

Figure 6.45 Simplified impedance characteristic of a wire with the matched impedance at far end.

using low-frequency excitation in an open circuit using the test setup in which far end of the measured wire is open. Similarly, the wire inductance L_C is a sum of partial inductances $L_C = l_C L_C'$ and can be measured using the test setup with far end of wire shorted to ground. Impedances of the wire in LF range for open and shorted far end can be approximated as purely capacitive and inductive, which results in an adequate impedance characteristics increase and decrease 20 dB per decade, respectively (Figure 6.45).

With the increase of frequency, the capacitive and inductive components of the wire impedance, dependent on the wire's per-unit parameters L_C' and C_C', are getting more and more close to each other and reach the same value above the characteristic transition frequency $f_{C,T}$. In the frequency range above the transition frequency impedances of the wire at opened and shorted configurations become the same and equal to the wire's characteristic impedance Z_{0C} determined according to the formula (6.24), if the far end of wire is loaded by the impedance equal to the characteristic impedance of the cable Z_{0C} (Figure 6.45).

In case when a load impedance connected to far end of the wire do not match the wire's characteristic impedance, load impedance differ from Z_{0C}, (in the utmost case a wire is open or shorted to the ground), signal reflections occur and standing waves are generated in the wire. The standing wave in the wire is a result of both the forward V_F and the reflected V_R waves.

Standing waves, generated in a wire treated as a TL, indicate the status of impedance matching and as a result measured voltages and currents appearing in a TL depend on the measuring point. Resulting resonant frequencies of wire are an effect of the velocity of signal transmission. The lowest, fundamental resonant frequency of wire occur when the quarter-wavelength $\lambda/4$ of transmitted signal is equal to the length of wire. Further resonant frequencies take place for every integer-multiple frequency of the fundamental resonant frequency associated to quarter-wavelength. It should be noticed that propagation speed of signals in real electrical wires is lower than the speed of light.

The signal propagation velocity is related to the relative dielectric permeability of cable insulation ϵ_r (6.25) and thus correlated with the propagation factor k_p, which can be determined based on the per-meter parameters of a wire according to the formula (6.26) where inductance L'_C and capacitance C'_C are equivalent per-unit length parameters [2].

$$v = \frac{1}{\sqrt{L'_C C'_C}} = \frac{1}{\sqrt{\mu_0 \varepsilon_0}} \cdot \frac{1}{\sqrt{\varepsilon_r}} = c_0 \cdot \frac{1}{\sqrt{\varepsilon_r}} \tag{6.25}$$

$$k_p(s) = \sqrt{(R'_C + sL'_C)(G'_C + sC'_C)} \tag{6.26}$$

Signal propagation velocity in power cables is usually approximately about two times lower than the light speed in vacuum c_0, because dielectric permittivity for typical cables insulations varies from 3 to 8, so the obtained velocity is usually within the range of 40–60% of the vacuum light speed. Considering the above, the signal wavelength can be determined according to the formula (6.27).

$$\lambda = v \cdot T == \frac{c_0}{\sqrt{\varepsilon_r}} \frac{1}{f} \tag{6.27}$$

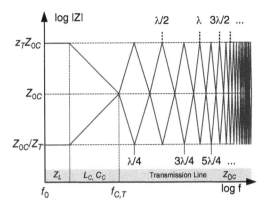

Figure 6.46 Simplified representation of oscillations of the wire impedance above characteristic transition frequency $f_{C,T}$ due to the wire resonances correlated to the wire length.

If reflection occurs at both ends of the wire, and physical length of the wire l_C perfectly fits to the integral multiple of half of wavelength λ of the transited signal, the wire works as a resonator and may cause large values of voltage and current to occur. Due to wire resonances occurring repetitively, at every frequency that makes the length of the TL equal to an integral multiple of 1/2 wavelength of the signal, TL input impedance also significantly changes in relation to the wire's characteristic impedance Z_{0C}. Therefore, the input impedance of the wire in the frequency range above $f_{C,T}$ (Figure 6.45) oscillates around the characteristic impedance Z_{0C} within the range from Z_{0C}/z_T to $Z_{0C}z_T$, where z_T is a normalized termination impedance factor (6.28).

$$z_T = \frac{Z_L}{Z_{0C}} \tag{6.28}$$

A simplified graphical representation of these recurring impedance changes, caused by the wire resonances, for the given unmatched load impedance $Z_L = z_T Z_{0C}$ is presented in Figure 6.46. Depending on the normalized termination impedance factor z_T, at frequency corresponding to $\lambda/2$, the lowest current or voltage resonance occurs, for $z_T > 1$ current and for $z_T < 1$, respectively. In both cases, the opposite resonances also occur for frequencies lower than $\lambda/4$ in relation to $n\lambda/2$. Resonating wire, due to generation of high currents and voltages, also works as a specific antenna, most effectively transmitting or receiving signals of frequencies corresponding to $\lambda/4$, $\lambda/2$, $3\lambda/4$, λ, $5\lambda/4$ ... .

In practice, HF losses in electrical wires working as a TL are noticeable, therefore, amplitudes of generated resonance currents and voltages are also noticeably decreasing with frequency. Considering losses in wires commonly

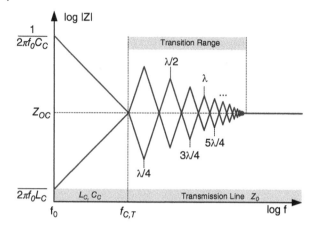

Figure 6.47 Simplified theoretical impedance characteristic of the power cable taking into account the effect of HF losses.

used in power circuits, the wide band impedance characteristic of open and shorted lines can be segmented into three subbands (Figure 6.47)–first: capacitive or inductive in the frequency range below $f_{C,T}$, second: transition range, above the frequency $f_{C,T}$ in which resonance effects are significantly visible, and the third: where resonance effects are damped and are not essential due to HF losses, thus exhibited impedance values are close to the characteristic impedance Z_{0C} of the cable.

Matching status of a TL is most commonly expressed by the reflection coefficient Γ, which is a vector quantity determined according to the formula (6.29), and for better understanding usually plotted on a complex plane inside the circle of radius 1. Magnitude of $\Gamma = 1$ means total reflection and $\Gamma = 0$ means no reflections.

$$\Gamma = \frac{V_R}{V_F} = \frac{Z_L - Z_{0C}}{Z_L + Z_{0C}} \tag{6.29}$$

The reflection coefficient Γ and the correlated normalized termination impedance z_T (6.30) are usually presented in a Smith chart form, which is a very efficient tool used for graphical development of impedance matching between the TL characteristic impedance $Z_{0,C}$ and impedance of the attached load Z_L (Figure 6.48).

$$z_T = \frac{Z_L}{Z_{0C}} = \frac{1 + \Gamma}{1 + \Gamma} \tag{6.30}$$

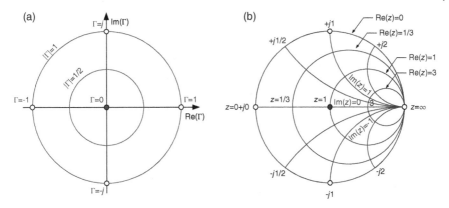

Figure 6.48 (a) Reflection coefficients presented on the complex plane as circles. (b) The impedance grid useful for graphically solving a impedance matching problem.

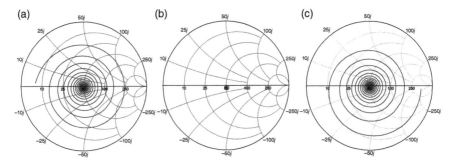

Figure 6.49 Smith chart representation of the complex reflection coefficient Γ for differently unmatched lossy wires, for selected termination impedances; $Z_T = 0.1$ (a), $Z_T \approx 1$ (b), $Z_T = 10$ (c).

Exemplary reflection characteristics of a lossy TL loaded by different load impedances $z_T = 1/5$, 1, and 5 are presented in Figure 6.49. Representation of a reflection characteristic on a Smith chart allows for comfortable assessment of matching status in the entire frequency range.

Localizations of the largest distances between the reflection characteristic and center of the Smith chart $|\Gamma|$ allow for simple recommendation of advantageous configurations of matching networks that can be applied for improvement of the signal transmission. Summarizing the considerations already presented, it can be found that the behavior of the wire operating as a signal transmission line depends not only on wire parameters but also on the frequency of the transmitted signal. For each TL, a specific transition frequency can be found, above which HF effects start to appear extensively. This transition frequency is tightly related to the frequency of the transmitted signal and the electrical length l_e

of the considered TL expressed in terms of the wavelength λ, wherein λ is the wavelength determined by considering signal propagation velocity (6.27) in the particular transmission line, dependent on its parameters. It should be noticed that the electrical length of TL l_e is defined for a given frequency of the transmitted signal, therefore, for transmission of broadband signals, a frequency of the highest harmonic component should be taken into account.

Estimation of the transition frequency $f_{C,T}$ of a TL can be explained based on simplified impedance characteristics of a cable presented in Figure 6.47. According to the presented simplification, the lowest resonance of a wire occurs at the frequency corresponding to $\lambda/4$, but the transition frequency $f_{C,T}$ can be roughly estimated as about two times lower frequency, that is, $\lambda/8$. It is the lowest frequency for which open- and short-circuit impedance characteristics achieve the same values, equal to the characteristic impedance Z_{0C} (6.31).

$$\lambda/8 \approx \frac{c_0}{\sqrt{\varepsilon_r}} \frac{1}{f_{C,T}} \tag{6.31}$$

In impedance characteristics of real wires, depending on losses factor, this frequency is usually slightly lower than $\lambda/8$; however, commonly $\lambda/10$ is assumed as length corresponding to the transition frequency $f_{C,T}$ of a wire, above which HF effects related to signal reflections begin to occur intensively. Therefore, to avoid HF effects in a TL, for the given frequency of the transmitted signal, the length of the line should be kept shorter than 10% of the wavelength of transmitted signal λ, or for the given TL length the frequency of transmitted signals should be not higher than $f_{C,T}$.

Unfortunately, in ASDs, in cable connection between the output side of FC and an AC motor, compliance to this rule is difficult to achieve, because of a wide spectrum of voltage harmonics generated due to rapid switching of voltage at the FC output. The spectrum bandwidth of the generated output voltage usually easily covers the entire bandwidth of conducted emission up to 30 MHz. Nevertheless, according to considerations presented in Chapter 2, magnitudes of generated harmonics start to decrease significantly above the frequency f_k correlated to the transistor switching time t_{on} (Figure 2.5). Thus, assuming significant decrease of voltage harmonic magnitudes above the knee frequency f_k, the switching time of transistor roughly allows estimating the upper limit of the spectral content of transmitted voltage, which can effectively initiate HF effects in transmission lines.

From the point of view of analysis of transmission of digital signals, characterized in time domain, a critical cable length can be estimated based on the rise/fall time of switched voltage t_{on}. The critical cable length occurs when the time needed for round-trip propagation of signal, from a source to load and back is equal to t_{on} (6.32), where t_d is a delay of signal propagation in wire.

$$t_{on} = 2t_d = 2\sqrt{L_C C_C} \qquad (6.32)$$

The critical wire length $l_{C,cr}$ resulting from voltage switching time can be calculated using formula (6.33) where v is the signal propagation velocity.

$$l_{C,cr} = \frac{1}{2}t_r \cdot v \qquad (6.33)$$

The estimated critical cable length $l_{C,cr}$ allows for rough assessment of possible level of exposure of the motor feeding cable to reflections phenomena of transmitted voltage, which is one of the most harmful effects for motor operating condition. Considering most commonly encountered switching times of IGBT transistors used in contemporary ASDs, mostly encountered distances between a motor and a FC from few to more than a hundred meters, and the frequency range of conducted emission up to 30 MHz, it can be noticed that TL effects can conceivably appear in the majority of ASD applications (Table 6.3). Nevertheless, a critical length of motor feeding cable can be a very useful indicator, which allows to estimate the intensity level of expected threats, which increase considerably when the critical cable length is exceeded.

6.5.3 Lumped Model of Shielded Power Cables

In ASDs, power cables with three working wires with a common shield that also perform a role of a grounding connection between a FC and a motor are usually used as a motor feeding cables [40]. It is also possible to use unshielded cables for motor feeding; however, this solution is rarely applied because of

Table 6.3 Short cable lengths estimated for characteristic frequencies associated to frequency range of conducted emission.

Frequency band	Frequency f	Wave length (in air) λ	Electrical length λ_e	Short cable length $l_e/10$
Power	50 Hz	6000 km	3000 km	< 300 km
Harmonics	2 kHz	150 km	75 km	< 7.5 km
HF Harmonics	9 kHz	33.3 km	16.6 km	< 1.6 km
CISPR A	150 kHz	2000 m	1000 m	< 100 m
CISPR B	30 MHz	10 m	5 m	< 0.5 m
Switching time 1 μ s	1 MHz	300 m	150 m	< 15 m
Switching time 0.1 μ s	10 MHz	30 m	15 m	< 1.5 m

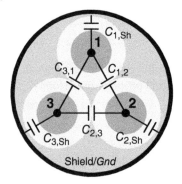

Figure 6.50 Equivalent parasitic capacitances of a three wire shielded cable with typical internal structure.

interference problems very likely occurring in nearby electrical systems. As it was described in previous sections, broadband behavior of unshielded cables in HF range is influenced by cable arrangement in relation to ground, which in real ASD installations is usually difficult to keep uniform all over the whole cable path. Therefore, to simplify presented considerations further analysis will be focused on the most frequently used three wire, shielded power cable with an internal structure presented in Figure 6.50.

Two types of parasitic capacitive couplings can be specified for three wire power cables: DM–between each of the two of the three wires, modeled by equivalent capacitances $C_{1,2}$, $C_{2,3}$, $C_{3,1}$ and CM–between each of the wires and the shield, modeled by capacitances $C_{1,Sh}$, $C_{2,Sh}$, $C_{3,Sh}$. Internal structure of a power cable is relatively more symmetrical in comparison to other three-phase components of an ASD, thus it can be assumed that the defined CM and DM equivalent capacitances associated with each wire are sufficiently similar to recognize them as equal. These capacitances can be readily estimated based on impedances of cable measurement at relatively low frequency, below the characteristic transition frequency $f_{C,T}$, where impedance characteristics are closely linear as illustrated on a logarithmic scale (6.45).

Referring to the considerations presented in previous sections, one of the most informative identification tests is measurement of cables' impedance characteristics with opened and shorted ends within an appropriate frequency range. This test is experimentally relatively simple and very useful for assessment of broadband behavior of cables. For multi wire cables, both the open and short tests can be completed for CM and DM signals components. The adequate test setup configurations for three wires, shielded motor cable are presented in Figure 6.51.

Four exemplary impedance characteristics acquired for 100 m long shielded cable are presented in Figure 6.52: $Z_{C,DM,OC}$–DM impedance in an open cir-

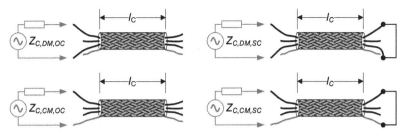

Figure 6.51 Test setups for measurement of CM and DM impedance characteristics of the motor cable.

cuit, $Z_{C,DM,SC}$–DM impedance in short circuit, $Z_{C,CM,OC}$–CM impedance in an open circuit, $Z_{C,CM,SC}$–CM impedance in short circuit. Supplementary dotted lines in Figure 6.52, tangential to the measured impedance characteristics of the cable at the identification of parameters frequency f_i, are added to improve the accuracy of estimation $C_{DM}(f_i)|_{@Z_{DM,OC,1-2}}$, $C_{DM}(f_i)|_{@Z_{CM,OC,1-Sh}}$ (dashed lines a and b) and inductances $L_{DM}(f_i)|_{Z_{DM,SC,1-2}}$, $L_{CM}(f_i)|_{Z_{CM,SC,1-Sh}}$ (dashed lines c and d), respectively, from short- and open-circuit impedance characteristics. The obtained cable's inductances can be directly converted into inductances per-unit length of the tested cable L'_{DM} and L'_{CM} by dividing their values by the length of the cable . Equivalent capacitances per unit length can be calculated based on measured capacitances $C(f_i)|_{@Z_{DM,OC,1-2}}$, $C(f_i)|_{@Z_{CM,OC,1-Sh}}$, according to formulas (6.34) and (6.35) resulting from the star–delta topology of DM and CM equivalent capacitances presented in Figure 6.50

$$C(f_i)\Big|_{@Z_{DM,1-2}} = C_{DM} + \frac{1}{2}\left(C_{CM} + C_{DM}\right) \tag{6.34}$$

$$C(f_i)\Big|_{@Z_{CM,1-Sh}} = C_{CM} + \frac{1}{2}\left(\frac{C_{CM}C_{DM}}{C_{CM} + C_{DM}}\right) \tag{6.35}$$

where $C_{DM,C} = C_{1-2} = C_{2-3} = C_{3-1}$ and $C_{CM,C} = C_{1-Sh} = C_{2-Sh} = C_{3-Sh}$.

Mutual inductances between the cable's wires have been omitted as they have less meaningful influence on CM conducted emission couplings [40]. The identified per-unit parameters, equivalent inductance L', and capacitance C' of the tested 100 m long shielded cable (3x2.5mm2) are collected in Table 6.4.

It should also be noticed that the measured open-circuit impedance characteristics of the tested cable in the frequency range below the transition frequency $f_{C,T}$ are sloped slightly less than 20 dB per decade, which is visible as slope differences between dashed lines a and b and the adequate characteristics.

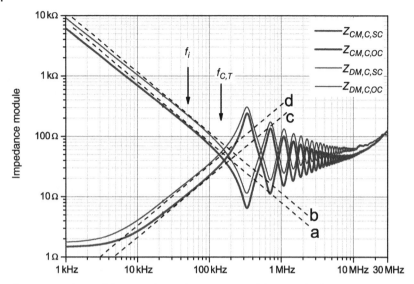

Figure 6.52 DM and CM impedance characteristics of the tested cable in open- and short-circuit testing configurations.

Table 6.4 Identified per unit parameters of the tested cable - 100m long, 3x2.5mm2, with braided shield

Mode	C'	L'	Z_{0C}
DM @ 50 kHz	150 pF/m	570 nH/m	62 Ω
CM @ 50 kHz	220 pF/m	350 nH/m	40 Ω

Such behavior is an effect of slightly decreasing parasitic equivalent capacitance caused by losses increasing with frequency and dependence of dielectric permittivity on frequency. Similar discrepancies between slopes of the measured impedance characteristics in short-circuit test configurations are effects of increasing conduction losses with frequency, resulting mainly form the skin and proximity effects occurring in cable wires.

6.5.4 Distributed Parameters Model of Shielded Power Cables

Impedance characteristics of power cables in the frequency range above the transition frequency $f_{C,T}$, as it is visible in Figure 6.53, expose reflection effects that result with repeatedly occurring resonances of voltage and current mode alternately. Frequencies of these resonances start from $f = 4c/\lambda$ and repeat every integer multiple of this value.

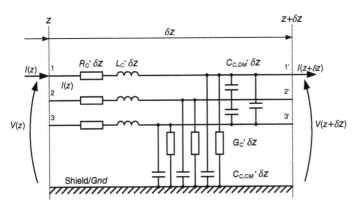

Figure 6.53 Lossy TL circuit model of three-wire-shielded power cable characterized by per-unit parameters: resistances $R_{C'}$ (Ω/m), inductances $L_{C'}$ (H/m), DM and CM capacitances $R_{C,DM'}$ and $R_{C,CM'}$ (F/m), and conductances $G_{C'}$ (S/m).

Values of parameters of the evaluated cable presented in Table 6.4 are different for DM and CM of transferred signals, which is a innate result of the cable structure presented in Figure 6.51. Nevertheless, frequencies of cable resonances $f_{\lambda/4}$, $f_{\lambda/2}$, $f_{3\lambda/4}$, f_{λ}, $f_{5\lambda/4}$, ... caused by signal reflections at cable ends are very similar for all measured impedance characteristics, as they are dependent on propagation velocity v (6.25) and physical length of a particular cable.

However, other parameters describing behavior of a cable in HF range, for example, the TL propagation factor (6.26), the characteristic impedance $Z_{0,C}$ (6.36) are different for DM and CM components of transmitted voltages and currents.

$$Z_{0C} = \sqrt{\frac{R'_C + sL'_C}{G'_C + sC'_C}} \approx \sqrt{\frac{L'_C}{C'_C}} \qquad (6.36)$$

Complete characterization of three-wire cables for DM and CM of signal transmission, in a frequency range within which TL effects occur and thus distributed character of the parameters must be taken into consideration, require identification of more per-unit parameters indicated in Figure 6.53. These parameters are usually substantially dependent on frequency, considering the frequency range of conducted emission.

Different propagation conditions of DM and CM voltage components at the three-phase output of FC and high levels of power transmitted within LF band enormously complicate simultaneous analysis of both of them in a wide-frequency range. Therefore, referring to the reasoning presented in Chapter 5, analysis of generation of conducted emission in ASDs, which are primarily CM phenomena initiated essentially in a motor feeding cable, might be simplified by separating analysis of HF CM currents and voltage components from DM components.

It can be assumed that during single commutation process in the output bridge of FC, transient processes in other two phases are usually already finished, as discussed Chapter 5, thus single transient accompanying to voltage switching in one phase can be analyzed separately, as the only HF process occurring at the particular time. Furthermore, representation during single voltage transient of the motor cable as only a single wire, coupled capacitively primary to the grounded shield of the cable and to other two wires, significantly simplifies analysis. The other two wires of the cable, during typical switching operation of FC output bridge, are connected to the DC bus voltage that usually exposes low impedance to the ground in HF range.

6.5.5 Identification of Broadband Parameters of Motor Feeding Cable

Generally, methods of analysis of cable connections are well developed and successfully used in applications where narrow-band transmission of HF signals is essential. Therefore, cables specially designed for use in these types of applications are comprehensively parametrized by manufactures. Power cables used as motor feeding cables in ASDs are designed primarily for power transfer at a low frequency close to the power line frequency. From the point of view of RF power transmission, these frequencies are very low. Transfer of power via power cables at frequencies significantly higher than power line harmonics above 2 kHz are very rarely considered as intentional RF power transfer because of low effectiveness. In most of the cases, transfer of power at HF via power cables is an undesired side effect that should be limited to avoiding subsequent side effects, like increased losses or interference. Motor feeding cable in ASD is one of an emblematic example of an application where HF voltage and current components are undesired but difficult to avoid. Therefore, parameters of power cables in HF range are commonly not specified by manufactures at all, or not precisely enough to perform analysis of conducted emission up to 30 MHz, thus have to be estimated individually based on experimental tests.

On the other hand, from the point of view of HF power transfer, power cables are specific because they must expose strong capability and effectiveness for power transfer at LF range, thus, for example, they have high-voltage isolation not optimized for RF, large cross-sections not optimized for skin and proximity effects, and so on. Therefore, not all RF engineering techniques used in HF bands can be directly applied for analysis of broadband power transfer, such as the one occurring at the output side of FC in an ASD. Furthermore, characteristic parameters of a power cable related to CM phenomena are especially rarely considered in design of most high-power applications, other then

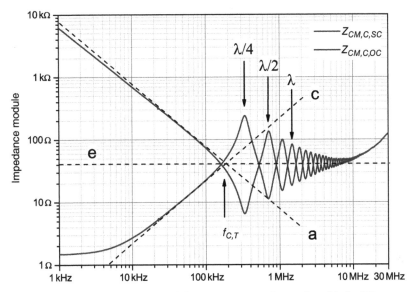

Figure 6.54 Graphical estimation of characteristic impedance of a cable for CM components of transferred signals.

ASDs, where broadband spectrum content is unintentionally transferred via long cable connections.

Fundamental parameter of a cable, from the point of view of behavior in HF range, is its characteristic impedance Z_{0C}, which is usually unspecified for power cables, especially for CM signals. The characteristic impedance $Z_{0C,CM}$ of the tested cable for CM signals can be estimated graphically, based on intersection points of open- and short-circuit impedance characteristics $Z_{CM,C,OC}$ and $Z_{CM,C,SC}$, in Figure 6.54 presented as dotted line e.

More accurate dependence of the characteristic impedance Z_{0C} of cable on frequency can be calculated according to formula (6.37).

$$Z_{0C} \approx \sqrt{Z_{C,OC} Z_{C,SC}} \tag{6.37}$$

The variability of characteristic impedances for the tested cable presented in Figure 6.55 shows that the characteristic impedances determined for DM and CM signals are not constant over the frequency range of conducted emission and substantially differ from each other. For the tested cable, the identified characteristic impedances for DM and CM signal are $Z_{0C,DM} \approx 60\,\Omega$ and $Z_{0C,DM} \approx 40\,\Omega$.

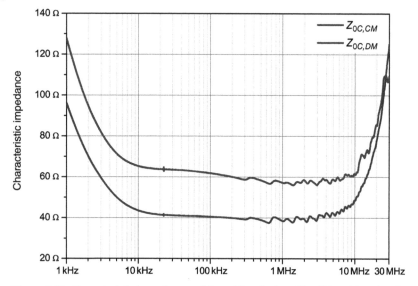

Figure 6.55 Characteristic impedances of the cable calculated for CM and DM signal components based on the impedance characteristics measured in open- and short-circuit configuration.

According to Figure 6.55, in the frequency range below 10 kHz, characteristic impedances of the tested cable are influenced by DC resistance of the cable. Between 10 kHz and few megahertz, slight decrease of the characteristic impedance with frequency is clearly observable. This is an effect of increasing losses and slight changes of the dielectric permittivity with the increase of frequency. The increasing part of impedance characteristic, above 10 MHz, can possibly be an effect of an extra inductance resulting from cable endings of the tested cable used for connecting cable to voltage source generator output, thus should be ignored as a "pig tail" effect. Furthermore, the impedance characteristics $Z_{CM,C,OC}$ and $Z_{CM,C,SC}$ indicate strong increase of HF losses in the cable with the increase of signal frequency. Increasing losses can be easily identified as significant decrease of impedance variations for subsequent frequencies related to resonances arising every $\lambda/4$ above the transition frequency $f_{C,T}$ (Figure 6.56). Due to considerably increasing losses, impedance characteristics of the cable tends relatively fast to the value of the characteristic impedance Z_{0C}. As a result, transmitted signals of frequencies significantly higher than $f_{C,T}$ after only few reflections, arising at frequencies corresponding to few wavelength λ, are strongly damped.

Based on analysis of impedance characteristics of the cable, it can also be noticed that the resonance frequencies caused by signal reflections repeating for subsequent frequencies resulting from wavelength $\lambda/4$ are not exactly equal to

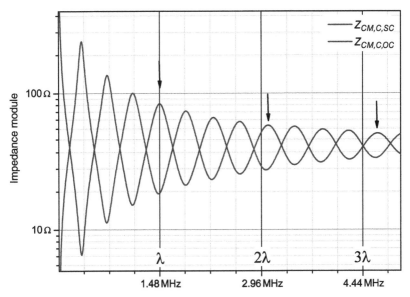

Figure 6.56 CM and DM impedance characteristics of a cable–resonances resulting from the electrical length of a cable and impedances mismatch.

multiples of $\lambda/4$ (Figure 6.54). In Figure 6.56, the frequency scale is referenced to frequency associated with wavelength λ, which was estimated as $f_\lambda \approx 1.48\,\text{MHz}$. According to the presented impedance characteristics, subsequent resonance expected at the frequency corresponding to 2λ actually occurs for slightly higher frequency than $f_{2\lambda} \approx 2.96\,\text{MHz}$. Similarly, the resonance expected at the frequency corresponding to 3λ is also higher than 4.44 MHz, and occurs at frequency shifted by the frequency offset that approximately corresponds to $\lambda/4$, which is significant change in relation to wavelength λ. These shifts of real resonance frequencies of the cable in relation to theoretically expected frequency are the result of changes in the equivalent parasitic inductances and capacitances, observed with the increase of frequency.

6.5.6 Broadband Modeling of Motor Feeding Cable

In a vast majority of ASD applications, a spectrum of voltages and currents transmitted between FC and controlled AC motor's windings is wide enough to initiate TL effects. Accurate modeling of TL effects occurring in motor feeding cables, which are not primarily designed for transmitting HF signals within such a wide-frequency range as covered by the conducted emission, is very difficult. Commonly, most of the parameters of motor cables are substantially dependent on frequency, especially in HF range where additional effects of

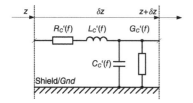

Figure 6.57 TL model of motor cable simplified to single wire representation adequate for analysis of single voltage transient generated during the FC output bridge commutation.

resonances occur. Accurate identification of these dependences is difficult to perform in a wide-frequency band.

On the other hand, very detailed approach to analysis of a cable behavior within the frequency range of conducted emission is not essential for analysis of CM currents generation in ASDs, as many other simplifications are necessary to apply. From the point of view of analysis of conducted emission in ASDs, two factors are most significant. The first factor is the values of resonance frequencies of the motor feeding cable that help to predict frequency subbands in which elevated conducted emission can occur. The second factor is level of HF losses in a cable that effectively reduces the effects of the resonances, especially those occurring in the frequency range well above the transition frequency $f_{C,T}$.

Resonance frequencies of the motor cable are directly dependent on signal propagation delay in the cable, which determines the electrical length of the cable. Experimental studies indicated that signal propagation velocity in the motor cable is also frequency dependent, because of dispersion phenomenon. This dependence is rather complex to model in details, but from the point of view of analysis of conducted emission can be represented by variable parasitic capacitances of cable wires. Changes of parasitic capacitances of cable wires with frequency is a result of frequency-dependent dielectric permeability of the cable insulation. Losses in cables, which commonly increase with frequency, are even more complex to model accurately, as they depend on more cable parameters, mostly associated with conduction losses in the wires and insulation material.

To model a variability of the motor cable DM and CM impedances more adequately in HF range using a simplified single wire TL model of the motor cable (Figure 6.57), two frequency-dependent per-unit parameters of a cable have been introduced: serial resistance $R'_C(f)$ and parallel parasitic capacitance $C'_C(f)$.

In the proposed frequency-dependent TL model of the motor cable, cable losses that increase in HF range are represented by frequency-dependent serial resistance $R'_C(f)$ (6.38). The effect of shifted resonance frequencies of the cable is modeled by frequency-dependent parasitic capacitance $C'_C(f)$ (6.38).

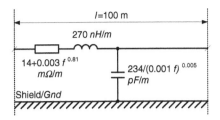

Figure 6.58 Experimentally identified parameters of a single wire TL model of the tested motor cable.

$$R'_C(f) = R'_{LF} + k_{R1} R'_{HF} f^{k_{R2}} \tag{6.38}$$

$$C'_C(f) = \frac{C'_{LF}}{(k_{C1}f)^{k_{C2}}} \tag{6.39}$$

It should be noted that such simplified representation of frequency-dependent behavior of a motor cable considered as a single wire TL, by introducing frequency-dependent variability of only two of four per-unit parameters, can be accepted only for analysis of cable impedances in frequency domain. More advanced analysis, especially in time domain, requires detailed identification of variability of all four per-unit parameters of the TL model of the cable. Modeling of variability of per-unit parameters with frequency is difficult to carry thoroughly within a wide-frequency range, especially in circuits in which multiple resonances occur. Therefore, for the purpose of analysis of conducted emission, coefficients of variability of the evaluated cable parameters k_{R1}, k_{R2}, k_{C1}, k_{C2} have been estimated based on the experimentally obtained impedance characteristics presented in Figure 6.54. Finally, obtained parameters of the complete, highly simplified TL model of the tested motor feeding cable are presented in Figure 6.58.

The proposed simplified approach to the motor cable modeling has been verified by comparison of the cable CM impedance characteristics obtained experimentally and calculated using the developed TL model (Figure 6.59). According to the presented comparison, resonance frequencies and increasing losses of the tested motor cable are reflected adequately enough up to the frequency of 10 MHz. A pig tail effect, associated with the cable connection ends used during impedance measurement, visible at the measured impedance characteristics above 10 MHz, can be modeled with an external inductance added to the TL model.

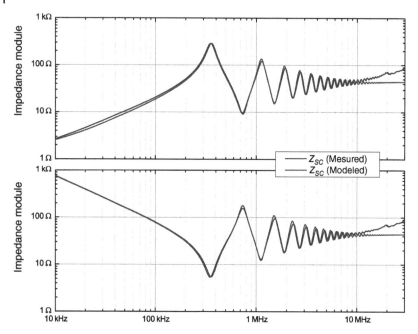

Figure 6.59 Comparison of the measured and modeled impedance characteristics of the tested motor cable.

6.6 Summary

Analysis of broadband behavior of high-power components used in ASD applications, like motor windings, motor feeding cable or inductors, is difficult because of parasitic couplings that in HF range usually become meaningful and have to be taken into consideration. Identification of circuit models' parameters of components of an ASD adequately representing their behavior in a wide-frequency range is very challenging, because accurate determination of parasitic couplings very often requires detailed specification of internal structure of these components. Broadband specifications of commercially accessible ASD components are usually not included in technical specifications commonly delivered by manufactures, for example, AC motors. Furthermore, most of the parameters related to broadband behavior of ASD components are usually significantly frequency dependent within the frequency range of conducted emission. Therefore, efficient modeling of generation and propagation of conducted emission in ASDs requires to apply rational simplifications that allow decreasing complexity of the developed models to a reasonable level. Considering the phenomena associated with CM currents generation at the output side of FC (which are the main origin of conducted emission in an ASD) highly simplified circuit models of the AC motor and the feeding cable are proposed. The main advantage of the proposed broadband circuit models is the possibility of identification of the models' parameters with satisfactory adequacy for conducted emission estimation, based on measurements of adequate impedances via externally accessible terminals only.

The proposed method of simplified analysis of CM currents generation at the output side of FC focused mainly on a single switching process of FC output bridge and representation of FC load as a single wire TL, therefore, allowing significant simplification of models and reduction of efforts needed for identification of model parameters. The developed simplified circuit models of AC motor windings and the feeding cable provide satisfactory mapping of CM impedance–frequency characteristics, thus allowing broadband modeling of CM currents distribution at the output side of FC within the conducted emission frequency range.

References

1 Y. Murata, K. Takahashi, T. Kanamoto, and M. Kubota, "Analysis of parasitic couplings in EMI filters and coupling reduction methods," *IEEE Transactions on Electromagnetic Compatibility*, vol. 59, no. 6, pp. 1880–1886, Dec 2017.

2 C. R. Paul, *Introduction to Electromagnetic Compatibility*. John Wiley & Sons, Inc., New York, 2006.

3 S. Wang, F. C. Lee, W. G. Odendaal, and J. D. van Wyk, "Improvement of EMI filter performance with parasitic coupling cancellation," *IEEE Transactions on Power Electronics*, vol. 20, no. 5, pp. 1221–1228, Sept 2005.

4 R. J. Kerkman, D. Leggate, D. W. Schlegel, and C. Winterhalter, "Effects of parasitics on the control of voltage source inverters," *IEEE Transactions on Power Electronics*, vol. 18, no. 1, pp. 140–150, Jan 2003.

5 J.-S. Lai, X. Huang, S. Chen, and T. W. Nehl, "EMI characterization and simulation with parasitic models for a low-voltage high-current AC motor drive," *IEEE Transactions on Industry Applications*, vol. 40, no. 1, pp. 178–185, 2004.

6 S. Wang, Y. Y. Maillet, F. Wang, R. Lai, F. Luo, and D. Boroyevich, "Parasitic effects of grounding paths on common-mode EMI filter's performance in power electronics systems," *IEEE Transactions on Industrial Electronics*, vol. 57, no. 9, pp. 3050–3059, Sept 2010.

7 J. Stewart, J. Neely, J. Delhotal, and J. Flicker, "DC link bus design for high frequency, high temperature converters," in *2017 IEEE Applied Power Electronics Conference and Exposition (APEC)*, March 2017, pp. 809–815.

8 H. Wang and F. Blaabjerg, "Reliability of capacitors for DC-link applications in power electronic converters: an overview," *IEEE Transactions on Industry Applications*, vol. 50, no. 5, pp. 3569–3578, Sept 2014.

9 A. Ayachit and M. K. Kazimierczuk, "Self-capacitance of single-layer inductors with separation between conductor turns," *IEEE Transactions on Electromagnetic Compatibility*, vol. 59, no. 5, pp. 1642–1645, Oct 2017.

10 G. Grandi, M. K. Kazimierczuk, A. Massarini, and U. Reggiani, "Stray capacitances of single-layer air-core inductors for high-frequency applications," in *Industry Applications Conference, 1996. Thirty-First IAS Annual Meeting, IAS '96., Conference Record of the 1996 IEEE*, vol. 3, Oct 1996, pp. 1384–1388.

11 J. L. Kotny, X. Margueron, and N. Idir, "High-frequency model of the coupled inductors used in EMI filters," *IEEE Transactions on Power Electronics*, vol. 27, no. 6, pp. 2805–2812, June 2012.

12 Z. D. GrØve, O. Deblecker, J. Lobry, and J. P. KØradec, "High-frequency multiwinding magnetic components: from numerical simulation to equivalent circuits with frequency-independent RL parameters," *IEEE Transactions on Magnetics*, vol. 50, no. 2, pp. 141–144, Feb 2014.

13 S. Racewicz, P. J. Chrzan, D. M. Riu, and N. M. Retiere, "Time domain simulations of synchronous generator modelled by half-order system," in *IECON 2012—38th Annual Conference on IEEE Industrial Electronics Society*, Oct 2012, pp. 2074–2079.

14 H. Zhang, S. Wang, D. Yuan, and X. Tao, "Double-ladder circuit model of transformer winding for frequency response analysis considering frequency-dependent losses," *IEEE Transactions on Magnetics*, vol. 51, no. 11, pp. 1–4, Nov 2015.

15 I. F. Kovacevic, T. Friedli, A. M. Musing, and J. W. Kolar, "3-D electromagnetic modeling of parasitics and mutual coupling in EMI filters," *IEEE Trans. actions on Power Electronics.*, vol. 29, no. 1, pp. 135–149, 2014.

16 I. F. KovaŁevi, T. Friedli, A. M. Msing, and J. W. Kolar, "Full PEEC modeling of EMI filter inductors in the frequency domain," *IEEE Transactions on Magnetics*, vol. 49, no. 10, pp. 5248–5256, Oct 2013.

17 M. Toudji, G. Parent, S. Duchesne, and P. Dular, "Determination of winding lumped parameter equivalent circuit by means of finite element method," *IEEE Transactions on Magnetics*, vol. 53, no. 6, pp. 1–4, June 2017.

18 B. Gustavsen and A. Semlyen, "Rational approximation of frequency domain responses by vector fitting," *IEEE Transactions on Power Delivery*, vol. 14, no. 3, pp. 1052–1061, Jul 1999.

19 Y. m. Zheng and Z. j. Wang, "Determining the broadband loss characteristics of power transformer based on measured transformer network functions and vector fitting method," *IEEE Transactions on Power Delivery*, vol. 28, no. 4, pp. 2456–2464, Oct 2013.

20 M. L. Heldwein, L. Dalessandro, and J. W. Kolar, "The three-phase common-mode inductor: modeling and design issues," *IEEE Transactions on Industrial Electronics*, vol. 58, no. 8, pp. 3264–3274, Aug 2011.

21 S. W. Pasko, M. K. Kazimierczuk, and B. Grzesik, "Self-capacitance of coupled toroidal inductors for EMI filters," *IEEE Transactions on Electromagnetic Compatibility*, vol. 57, no. 2, pp. 216–223, April 2015.

22 A. Ahmad and P. Auriol, "Dielectric losses in power transformers under high frequency transients," in *Conference Record of the 1992 IEEE International Symposium on Electrical Insulation*, Jun. 1992, pp. 460–463.

23 Y. Ryu and K. J. Han, "Improved transmission line model of the stator winding structure of an AC motor considering high-frequency conductor and dielectric effects," in *2017 IEEE International Electric Machines and Drives Conference (IEMDC)*, May 2017, pp. 1–6.

24 M. Eslamian and B. Vahidi, "New equivalent circuit of transformer winding for the calculation of resonance transients considering frequency-dependent losses," *IEEE Transactions on Power Delivery*, vol. 30, no. 4, pp. 1743–1751, Aug 2015.

25 Y. Sato, T. Shimotani, and H. Igarashi, "Synthesis of cauer-equivalent circuit based on model order reduction considering nonlinear magnetic property," *IEEE Transactions on Magnetics*, vol. 53, no. 6, pp. 1–4, June 2017.

26 M. M. Shabestary, A. J. Ghanizadeh, G. B. Gharehpetian, and M. Agha-Mirsalim, "Ladder network parameters determination considering nondominant resonances of the transformer winding," *IEEE Transactions on Power Delivery*, vol. 29, no. 1, pp. 108–117, Feb 2014.

27 J. H. Krah, "Optimum discretization of a physical cauer circuit," *IEEE Transactions on Magnetics*, vol. 41, no. 5, pp. 1444–1447, May 2005.

28 A. Boglietti, A. Cavagnino, and M. Lazzari, "Experimental high-frequency parameter identification of ac electrical motors," *IEEE Transactions on Industry Applications*, vol. 43, no. 1, pp. 23–29, Jan 2007.

29 J. Luszcz, "AC motor feeding cable consequences on EMC performance of ASD," in *2013 IEEE International Symposium on Electromagnetic Compatibility (EMC),*, 2013, pp. 248–252.

30 C. Petrarca, A. Maffucci, V. Tucci, and M. Vitelli, "Analysis of the voltage distribution in a motor stator winding subjected to steep-fronted surge voltages by means of a multiconductor lossy transmission line model," *IEEE Transactions on Energy Conversion*, vol. 19, no. 1, pp. 7–17, March 2004.

31 S. P. Weber, E. Hoene, S. Guttowski, W. John, and H. Reichl, "Modeling induction machines for EMC-analysis," in *2004 IEEE 35th Annual Power Electronics Specialists Conference*, vol. 1, June 2004, pp. 94–98.

32 A. Boglietti, A. Cavagnino, and M. Lazzari, "Experimental high frequency parameter identification of AC electrical motors," in 2005 IEEE International Conference on Electric Machines and Drives, 2005, pp. 5–10.

33 M. Moreau, N. Idir, and P. L. Moigne, "Modeling of conducted EMI in adjustable speed drives," *IEEE Transactions on Electromagnetic Compatibility*, vol. 51, no. 3, pp. 665–672, Aug 2009.

34 J. Sun and L. Xing, "Parameterization of three-phase electric machine models for EMI simulation," *IEEE Transactions on Power Electronics*, vol. 29, no. 1, pp. 36–41, Jan 2014.

35 N. Djukic, L. Encica, and J. J. H. Paulides, "Overview of capacitive couplings in windings," in *2015 Tenth International Conference on Ecological Vehicles and Renewable Energies (EVER)*, March 2015, pp. 1–11.

36 Q. Liu, F. Wang, and D. Boroyevich, "Conducted EMI noise prediction and characterization for multi-phase-leg converters based on modular-terminal-behavioral (MTB) equivalent EMI noise source model," in *37th IEEE Power Electronics Specialists Conference, 2006. PESC '06*, June 2006, pp. 1–7.

37 J. Luszcz, "Modeling of common mode currents induced by motor cable in converter fed AC motor drives," in *2011 IEEE International Symposium on Electromagnetic Compatibility (EMC)*. IEEE, 2011, pp. 459–464.

38 E. J. Bartolucci and B. H. Finke, "Cable design for PWM variable-speed AC drives," *IEEE Transactions on Industry Applications,*, vol. 37, no. 2, pp. 415–422, 2001.

39 N. Hanigovszki, J. Landkildehus, G. Spiazzi, and F. Blaabjerg, "An EMC evaluation of the use of unshielded motor cables in AC adjustable speed drive applications," *IEEE Trans. actions on Power Electronics.*, vol. 21, no. 1, pp. 273–281, 2006.

40 G. Skibinski, R. Tallam, R. Reese, B. Buchholz, and R. Lukaszewski, "Common mode and differential mode analysis of three phase cables for PWM AC drives," in *Conference Record of the 2006 IEEE Industry Applications Conference Forty-First IAS Annual Meeting*, vol. 2, Oct 2006, pp. 880–888.

7

Impact of Motor Feeding Cable on CM Currents Generated in ASD

The only source of knowledge is experience.
Albert Einstein

The most considerable part of CM currents generated in ASDs is initiated at the output side of FC during voltage commutations. The motor feeding cable is usually connected directly to FC output terminals or sometimes via passive filters with simple topologies, L or LC. Therefore, parasitic parameters of the motor feeding cable play a key role in CM currents generation and propagation toward the AC motor and the power grid.

Broadband characteristics of a motor cable, as the most broadly excited component of an ASD, powerfully influence magnitudes and spectral distribution of CM currents between the motor, FC, and the power grid. CM currents injected into AC motor windings, transmitted via DC bus of FC to the grid side of FC and also collected in grounding connection of FC, can be significantly influenced due to broadband properties of the motor feeding cable.

Broadband characteristics of three-wire power cables commonly used for feeding AC motors in ASD have a relatively regular and balanced structure in comparison to other ASD components, for example, FC or motor windings. Nevertheless, the length of motor feeding cable, which in each case is individually matched for a particular application, can significantly change the overall performance of an ASD in terms of conducted emission generation. Moreover, the motor feeding cable should meet requirements related to transfer of high power at low frequencies; therefore, optimization of its performance in HF range is additionally limited.

Various recommendations for motor feeding cables used in ASDs have been developed by manufacturers. Those recommendations help to avoid excessive EMC problems; nevertheless, the influence of the motor feeding cable on overall

High Frequency Conducted Emission in AC Motor Drives Fed by Frequency Converters: Sources and Propagation Paths, First Edition. Jaroslaw Luszcz.
© 2018 by The Institute of Electrical and Electronic Engineers, Inc. Published 2018 by John Wiley & Sons, Inc.

EMC performance of an ASD can still be crucial in many of contemporary applications. Accurate modeling of HF phenomena occurring at the output side of FC is very difficult because of omnipresent parasitic couplings that become meaningful in the frequency range of conducted emission. The method of simplified analysis of broadband behavior of FC's load in the frequency range of conducted emission, presented in this chapter, allows recognizing fundamental effects of influence of motor feeding cable on CM currents generation and distribution in ASD [1–3].

7.1 Influence of Motor Feeding Cable on CM Impedance of FC's Load

Generally, from the point of view of broadband CM signals transmission, circuit representation of the motor feeding cable can be simplified to a π-network, assuming that only simplified analysis of one of the wires during single voltage switching is considered (Figure 7.1). This means that general model of the motor cable can be represented by transverse shunt impedances $Z_{MC,Par}$ and longitudinal serial impedances $Z_{MC,Ser}$, that are uniformly distributed along the cable and are greatly frequency dependent.

Equivalent shunt impedances $Z_{MC,Par}$ are associated with CM parasitic capacitances of cable wires in relation to the grounded cable's shield and dielectric losses. These distributed shunt impedances of the motor feeding cable form an extra sub loop $I_{CM,MC}$ for circulation of CM currents generated at the FC output $I_{CM,Out}$, which are initiated by the motor feeding cable exclusively. Equivalent longitudinal impedances of the motor cable $Z_{MC,Ser}$ are associated with cable's parasitic inductances and conduction losses, which result in

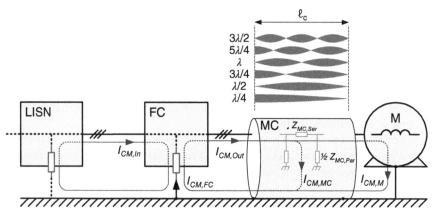

Figure 7.1 Common mode currents' loops in ASD with shielded motor cable.

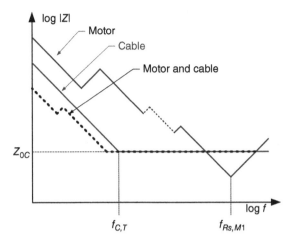

Figure 7.2 Simplified graphical estimation of the CM impedance characteristic of the motor with the feeding cable based on CM impedance characteristics of the motor and the cable.

transmitted voltage spectra via motor feeding cable are also attenuated by the cable, accordingly to ratios of the π-network model impedances of the cable (Figure 7.1).

Lumped component circuit models of the motor cable can be effectively used in frequency range only below the transition frequency of the cable $f_{C,T}$, which is closely dependent on its length (Figure 6.47). In the frequency range above the cable's transition frequency $f_{C,T}$, the motor feeding cable behaves as a transmission line with the characteristic impedance $Z_{0,C}$ dependent on parasitic parameters of the cable, both capacitances and inductances. Due to signal reflections, which in ASDs usually take place at both ends of the motor feeding cable, standing waves can be generated at specific frequencies correlated to multiples of $\lambda/4$.

Therefore, insertion of motor feeding cables of different lengths between the FC output and motor windings significantly change the resultant impedance of FC load seen via the FC's output terminals. General principle of influence of the motor feeding cable on modification of the CM impedance of the FC load is presented in Figure 7.2.

The CM impedance of the motor cable, for typically used lengths of motor feeding cables, is usually of similar order or smaller than the CM impedance of motor windings in the frequency range below $f_{C,T}$, whereas the characteristic impedance of cable $Z_{0,C}$, characterizing cable behavior in the frequency range above $f_{C,T}$, is usually higher than the lowest CM impedance of motors windings occurring at the frequency band close to the main serial resonance $f_{Rs,M}$ (Figure 6.30).

In the frequency range below $f_{C,T}$, the insertion of the motor feeding cable causes a decrease of the resultant CM impedance of the converter's load in relation to the CM impedance of the motor only. However, in the frequency range above $f_{C,T}$, this relation changes, the resultant CM impedance of the motor with the feeding cable becomes close to the CM impedance of the cable only, as the characteristic impedance of the cable is affecting more the resultant impedance of the motor with a cable. Therefore, in the frequency subband where the CM impedance of motor windings is the lowest, close to the frequency of the main serial resonance $f_{Rs,M}$, the effect of the motor cable on the resultant CM impedance of the FC load can be opposite, that is, it can increase instead of decreasing. For the given simplified CM impedance characteristics of the motor and the cable presented in Figure 7.2, the resultant CM impedance characteristic of motor with cable is presented as a dotted line. It should be noted that in such a highly simplified representation of CM impedance characteristics, voltage reflection effects in a cable are not taken into account to emphasize the role of the characteristic impedance of cable $Z_{0,C}$.

Summarizing, the CM impedance of the FC's load seen by the output-side terminals of FC, which is essential for generation of CM currents, is significantly influenced by the motor feeding cable, especially by its length. CM impedance changes caused by the motor cable are strongly dependent on frequency and can be remarkably different. The increase or decrease of the resultant FC load impedance can be observed, depending on the frequency subranges defined by frequencies of resonances occurring on individual CM impedance characteristics of the cable and the motor.

Analysis of influence of motor feeding cables on CM impedances of FC's loads require to take TL effects into consideration, and associated standing waves, which occur in the specific frequency ranges, depending on cable properties (mainly its length). Therefore, the influence of the length of the motor feeding cable on the CM impedance seen at the output side of FC can be estimated only individually for each particular application, for which wide-band impedance characteristics of the motor are known. In other words, the influence of the same cable can be significantly different for different motors [4].

7.2 Broadband Modeling of Motor Cable Impact on CM Currents Generated in ASD

There are a number of known adverse side effects of CM currents flowing in different components of an ASD, which should be reduced for better performance of an ASD itself and for protection of other neighboring systems against possible interference caused by conducted and radiated emission [5]. CM currents generated in ASDs are mostly initiated at the output side of FC, and are closely correlated to the CM impedance of FC load, which commonly consists of the motor windings and the attached feeding cable. Therefore, CM impedance characteristics of individual components of an ASD, especially connected to the output side of FC, discussed in previous chapters are a preliminary fundamental stage that allows further more comprehensive investigation of CM currents distribution in ASD applications.

According to the essential CM current loops existing in the ASD, presented in Figure 7.2, three key measuring points of CM currents, mostly important from the point of view of CM currents generation, can be indicated: first—AC motor, second—FC output, and third—the power grid. The motor feeding cable affects all these defined CM currents: $I_{CM,M}$, $I_{CM,Out}$, $I_{CM,In}$, and additionally introduces one more CM current component $I_{CM,MC}$, flowing through the motor cable only, which is a part of total CM current generated at the output side of FC $I_{CM,Out}$.

A wide spectrum of the investigated frequencies and resonance effects possible to occur cause that the influence of the motor feeding cable on magnitudes of generated CM currents and their distribution in ASDs is difficult to predict in both senses, in a sense of the influence on the magnitudes and in a sense of its character, positive or negative. Complex networks of parasitic couplings are especially difficult to identify in three-phase systems and a wide range of magnitudes of the analyzed signals cause that modeling of CM currents in ASDs in time domain often not effective [6–8].

Referring to simplified wide-band circuit models of ASD sub components presented in Chapter 6, analysis in frequency domain is proposed for assessment of the influence of the motor feeding cable on CM currents generation and propagation in an ASD. The N-port network approach, used at the components identification stage, allows further cascade connections of these models into expanded models covering greater part of an ASD.

Accepting the previously adopted assumption that the level of symmetry between individual phases in impedances of FC load is high enough, single-wire representation of an ASD can be used instead of a full three-phase circuit model. This assumption results in that the proposed single-wire approach more adequately represents behavior of the output side of FC during generation of a single voltage transient in only one phase of an ASD in a given time. Therefore, DM phenomena accompanying the output voltage commutations simultaneously are not possible to be reproduced using this single-wire approach. However, one-wire approach in modeling of CM currents allows significant simplifications of the proposed models of an ASD and also strongly reduces workloads necessary for wide-band identification of parameters of these models. Furthermore, one-wire approach emphasizes effects of resonances associated with CM impedances, which are primarily essential for conducted emission estimation.

A circuit model of a complete inverter's load composed of two sub-models, a TL model of the motor feeding cable and a ladder circuit model of the motor

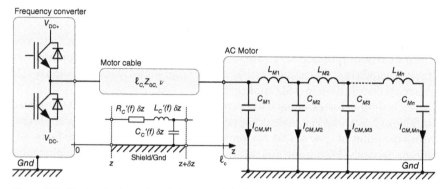

Figure 7.3 Single-wire broadband circuit model of the complete load of FC designed for CM currents analysis in frequency domain.

windings representing distributed parasitic capacitive couplings, is presented in Figure 7.3. A cascade structure of the proposed model allows easy modification of cable length l_C and thus simulation study of its influence on magnitudes of CM currents generated at the output side of FC. Furthermore, such a cascaded circuit model allows investigation of CM currents distribution between the motor feeding cable and the windings, which is dependent on the cable length and particularly related with the impedance mismatch coefficient occurring at the motor's terminals, between the cable and the windings.

7.2.1 CM Impedance of Frequency Converters' Load

Results of an example simulation analysis of influence of the motor feeding cable length on the CM impedance seen by the FC's output terminals for the evaluated AC motor are presented in Figure 7.4. On the basis of the obtained results, the most meaningful changes of the CM impedance characteristic of FC's load introduced by feeding cables of different length can be characterized as follows:

- A decrease of the CM impedance in the frequency range up to the frequency of lowest resonance of the attached cable associated with a quarter of electrical length of the cable $\lambda/4$. This decrease, indicated by the behavior marked with the arrow A, closely corresponds in size to the parasitic CM capacitance of cable wires, determined in frequency range below the cable transition frequency $f_{C,T}$, within which distributed character of cable parasitic CM capacitances is not essential and can be neglected.
- A decrease of frequency of the load parallel resonance correlated to the main parallel resonance of the motor windings $f_{Rp,M}$ (indicated by the arrow B), which for the evaluated motor has been observed between 30–40 kHz. For the evaluated case, for a 300 m long motor feeding cable, this resonance

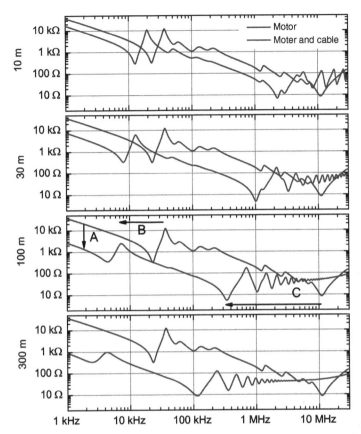

Figure 7.4 CM impedance characteristics of motor with feeding cables of different lengths in comparison to CM impedance of same motor windings.

frequency of FC's load decreases a factor of about 10 in relation to the corresponding resonance frequency of the motor windings without feeding cable.

- A decrease of frequency of the load serial resonance correlated to the main serial resonance of the motor windings $f_{Rs,M}$ (indicated by the arrow C), which for the evaluated motor has been observed at the frequency just above 1 MHz. For the evaluated case, for a 300 m long motor feeding cable, this frequency decreases about 100 factor in relation to the corresponding resonance frequency of the motor windings without a feeding cable.

When comparing CM impedances of the motor with feeding cable and the cable only, it can be noticed that the CM impedance of FC's load strongly tends to the impedance of the cable only with the increase of the cable length (Figure 7.5). For the evaluated case, for the length of the motor feeding cable of

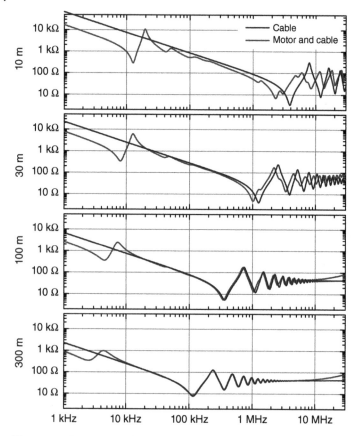

Figure 7.5 Comparison of calculated CM impedances of the motor and the motor with feeding cables of different lengths connected.

100 m, the CM impedance of the motor with feeding cable is nearly equal to the impedance of the feeding cable within the entire frequency band of conducted emission, 9 kHz to 30 MHz. This means that CM currents generated at the output side of FC are mainly dependent on cable parameters for cables longer than several tens of meters.

The positive effect of this coincidence is that the minimal CM impedance of motor windings, occurring at the frequency of main serial resonance of windings $f_{Rs,M}$, which is usually much higher than the cable transition frequency $f_{C,T}$, is increased by added motor feeding cable to the level of the characteristic impedance of cable Z_{0C}, which is usually higher than winding impedance in this frequency range. For the evaluated case, this phenomenon can be observed for frequencies of about 1 MHz (Figure 7.3), for which the motor impedance

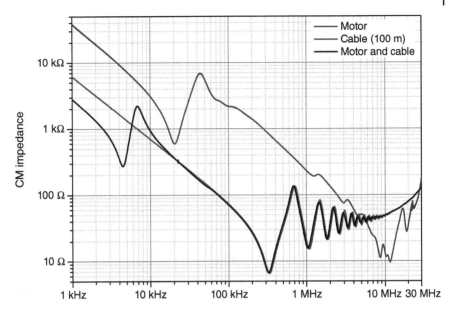

Figure 7.6 Comparison of measured CM impedances: of the motor windings, 100 m-long motor feeding cable, and the motor windings with feeding cable.

lower than 10 Ω has been increased up to about 40 Ω, which is very close to the characteristic impedance of the used cable.

Experimental verification of the presented method of analysis of influence of the motor feeding cable on the resultant CM impedance of FC's load has been done for the evaluated motor and 100 m-long shielded cable. Corresponding measured CM impedance characteristics of the motor windings, the feeding cable, and the motor with feeding cable are presented in Figure 7.6. Taking into consideration the level of adopted simplifications in modeling of the evaluated system, presented simulation results in Figures 7.4 and 7.5 exhibit good agreement with experimentally obtained CM impedances presented in Figure 7.6.

7.2.2 Reflections at Motor's Terminals

The motor feeding cable is a key component of an ASD that behaves as a TL for HF signals transmitted over adequately long distance (Section 6.5). Efficiency of transmission of HF signals via motor feeding cable characterized mainly by the characteristic impedance of the cable Z_{0C} determined for CM signal components depends on matching impedances seen at both ends of the cable. One end of the motor feeding cable is connected to the FC output, which is a source of transmitted signals, and the second end to the AC motor windings, which play the role of load for a TL formed by the motor cable (Figure 7.7).

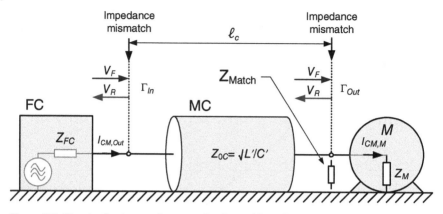

Figure 7.7 Signal reflections at the motor feeding cable ends.

The output impedance of FC as a signal source is difficult to determine, because it is influenced by switching processes occurring at the output bridge of FC. The impedance of a cable load, which is the motor windings, is constant in time, although it is strongly dependent on frequency because of parasitic couplings occurring in motor windings. Therefore, broadband matching of impedances at the motor cable ends is a very troublesome task, difficult to solve efficiently in a wide frequency range.

From the point of view of generation of CM currents in ASDs, signal reflections occurring at the motor side of the feeding cable are the most meaningful [9–12]. The reflection coefficient Γ at the windings terminals can be calculated based on the CM impedance characteristic of motor windings according to Eq. 6.29. A reflection characteristic calculated within the frequency range of conducted emission for the evaluated AC motor and feeding cable is presented in Figure 7.8.

Based on the calculated reflection characteristic, it can be noticed that signals of frequencies below 1 MHz are strongly reflected at the motor terminals, excluding a narrow frequency range close to the frequency of about 20 kHz, where a serial resonance in the motor windings exists. This means that transmission of signals in this frequency range will result with an increase of magnitudes of voltages occurring at the motor terminals well above the level resulting from the DC bus voltage value. Over-voltages caused by voltage reflections at the motor's terminals are very harmful for motor windings insulation, but can also significantly change conditions of generation of CM currents in motor windings.

The frequency characteristic of reflection magnitudes $|\Gamma(f)|$ of the evaluated motor cable presented in Figure 7.8 allows advantageous identification of frequency sub bands within which the reflections of transmitted signals are

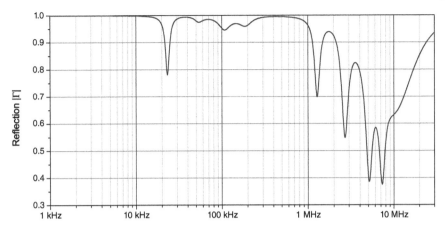

Figure 7.8 Reflection characteristic $|\Gamma_{Out}|(f)$ at winding terminals of the evaluated AC motor.

most critical. The same reflection coefficient represented as vector quantity in a form of a Smith chart is very helpful for developing matching networks that can be inserted at the motor terminals to achieve better matching coefficient characteristic of winding impedance within as broad frequency range as possible. Smith chart characteristic of complex reflection coefficient Γ_{Out} of the evaluated motor is presented in Figure 7.9.

Depending on the frequency range in which a matching impedance is expected to be the most effective, by using Smith chart representation of the complex reflection coefficient, an efficient configuration of a matching network Z_{Match} can be found (Figure 7.7). Exemplary results of simulation analysis of impedance matching at the motor's terminals have been used for the matching impedance Z_{Match} consisting of the resistance equal to the characteristic impedance of the motor feeding cable Z_{0C} connected in series with a capacitance that increases the impedance of the attached matching RC network in a low frequency range [9].

Effects of impedance matching at the motor terminals on a reflection characteristic can be clearly presented in a Smith chart (Figure 7.10). It can be seen that the reflection coefficient is significantly reduced in HF range, and only for low frequencies remains higher than 0.5. Nevertheless, from the point of view of the generated conducted emission, CM currents generated in LF range are much lower, and thus less meaningful.

CM currents generated at the output side of FC as a key indicator of conducted emission can also be determined using the proposed simplified circuit model. More detailed frequency characteristics of influences of applied matching network on CM currents generated at the FC output $I_{CM,Out}$ and in the motor windings $I_{CM,M}$ are presented in Figure 7.11. The obtained simulation

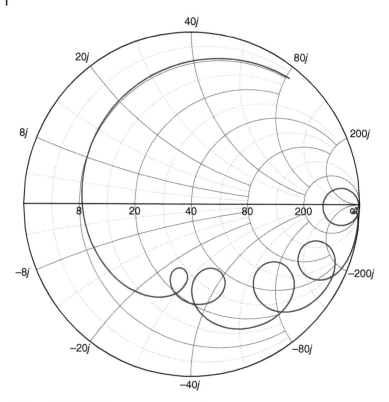

Figure 7.9 Smith chart representation of complex reflection coefficient Γ_{Out} at terminals of the evaluated motor.

results are presented as relative characteristics, in which the level of 100% is the reference level of CM currents occurring in an ASD without matching components applied.

Based on the presented results of simulation analysis, it can be noticed that the use of impedance matching RC networks causes the CM currents generated at the FC output $I_{CM,Out}$ significantly increase in the frequency range below the transition frequency of the cable $f_{C,T}$. In a higher frequency range, regularly repeating effects of the matching network are observed, correlated to frequencies of cable resonances caused by signal reflections in the cable. The CM currents emission characteristic in this frequency range is significantly flattened, in such a way that peak CM current values occurring at frequencies correlated to multiples of quarter of the electrical cable length ($\lambda/4$) are decreased, whereas minimum values occurring at frequencies correlated to multiples of half of the electrical cable length ($\lambda/2$) are increased.

(a)

(b)

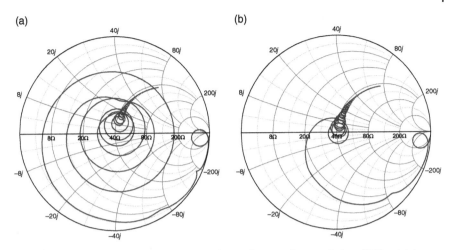

Figure 7.10 Influence of a matching network on reflection characteristics of FC load. (a) Without matching. (b) With an RC matching network.

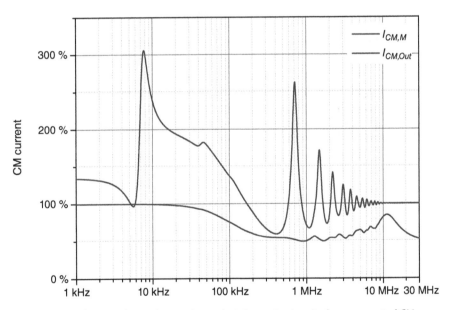

Figure 7.11 Influence of impedance mismatch at the motor terminals on generated CM currents at the output side of FC and in the motor windings.

Influential effects of applied matching RC networks on CM currents in the motor windings $I_{CM,M}$ (Figure 7.11) can be observed in the most significant frequency range in which cable-originated resonances occur, that is, above the transition frequency of the cable $f_{C,T}$. Nevertheless, this influence is not large, in the evaluated case up to about two times; therefore, the cable impedance matching method is primarily used for reducing transient over voltages caused by reflections occurring at motors' terminals [13, 14]. The obtained slight decrease of generated CM currents in motor windings is often only a supplementary positive effect of impedance matching. It should be also noted that power losses dissipated at termination networks that include resistances that can be significantly high and are one of substantial disadvantages of this method.

7.3 Motor Cable Influence on CM Currents Distribution in ASD

CM currents, as one of the main indicators characterizing conducted emission generated in ASDs, are mainly dependent on spectral content of voltages generated at the FC output terminals and on properties of attached load characterized by CM impedances. As the resultant CM currents generated in ASDs are also dependent on many other parameters, the impact of the motor feeding cable has been analyzed comparatively, to extract more clearly the effect of the evaluated motor feeding cable only. The influence of the motor cable on CM currents generated in an ASD has been defined as a ratio between the reference emission characteristics obtained in the application with a very short cable and the same application but with significantly greater cable lengths, which is the most varying and meaningful parameter of motor feeding cables.

Changes of CM impedance characteristic of inverter's load as a result of attached motor feeding cables of different lengths are the main cause of modification of generated spectra of CM currents in final applications. In the investigations, the impact of the motor feeding cable on CM currents generated in an ASD has been examined toward CM current changes in three characteristic measuring points of ASD:

- At the inverter's output that increase conducted and radiated emission of the ASD.
- At the motor winding terminals that can be harmful for AC motor itself and increase its failure rate.
- At the grid side of FC that is required according to the limits defined in EMC-related standards [15, 16].

As spectral characteristics of the FC output voltage and CM impedances of the FC load are frequency dependent, the spectral characteristic of generated CM currents can also be strongly influenced by mutual location in frequency domain of resonance frequencies occurring in the generated voltage spectrum (Figure 2.5) and the CM impedance characteristics of the FC load (Figure 7.6).

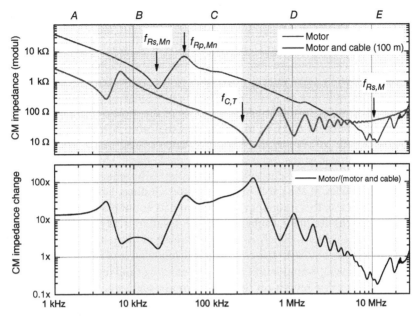

Figure 7.12 The rate of change of CM impedance of the FC load caused by the motor cable: measurement data for the evaluated motor and 100 m-long feeding cable.

7.3.1 CM Currents at Frequency Converter Output

CM current generated at the output side of FC $I_{CM,Out}$ directly depends on CM impedance of inverter's load and generally increases with the length of the cable. Nevertheless, due to existing multiple resonance effects, this dependence can be very diversified in different frequency ranges. Direct comparison of CM impedances of motor windings alone and the motor windings with feeding cable allows rough estimation of the cable influence on CM currents generated at the output side of FC (Figure 7.12).

Based on the comparison of CM impedances of the motor alone and the motor with feeding cable, characteristic frequency subranges A, B, C, D, and E, associated with resonance frequencies of the motor windings $f_{Rs,M}$, $f_{Rs,Mn}$, $f_{Rp,Mn}$ and the transition frequency of the cable $f_{C,T}$ can be recognized. In the frequency subranges A and C, relatively constant decrease of the CM impedance is observed as there are no significant resonance effects to occur. These impedance changes are correlated to the cable wires' CM parasitic capacitances relatively easily determinable in the frequency range below $f_{C,T}$. In the frequency subrange B, where the motor's main parallel resonances typically occur in the AC motor's winding (Figure 6.29), resonance effects have significant influence; thus, in this range prediction of the resultant CM impedance of the motor with the cable is more difficult. In this frequency range, resultant

CM impedance characteristics can be very sensitive even to little changes of load parameters.

In the frequency subrange D, resonance effects of signal reflections at the motor feeding cable ends are visible; thus, impedance characteristic is changing significantly, but in a certain way on a regular basis, respectively to frequencies correlated to multiples of $\lambda/4$. In this frequency range, with the increase of the cable length, the resultant CM impedance of the motor with feeding cable tends to be similar to the CM impedance of the cable only (Figure 7.5). For the cable length of 100 m, the CM impedance of the motor with cable is already very similar to the impedance of the cable itself. Thus, the amplifying effect of the cable for CM currents is decreasing with frequency up to that for which impedances of the cable and the motor windings become similar. For the evaluated case, this frequency was detected as about 5.5 MHz. Above this frequency, in the frequency subrange E, the resultant CM impedance of the motor with the cable is usually increased in relation to the CM impedance of the motor and very close to the characteristic impedance of the cable $Z_{0,C}$.

Using the developed models of the motor windings and feeding cable, wider analysis of the cable length on the CM currents spectrum at the FC output has been done for cable lengths from 10 to 300 m. Based on the results of these analyses that are presented in Figure 7.13, it can be figured out that the frequency subrange marked as C becomes narrower with the increase of the cable length, and simultaneously the frequency range of subrange marked as D expands.

Values of the transition frequency between the subranges C and D are closely correlated to transition frequencies $f_{C,T}$ of cables of different lengths, starting from about 4 MHz for a 10 m long cable up to about 100 kHz for a 300 m long cable. In the whole frequency subrange C, increase of CM currents at the FC output caused by the feeding cable corresponds to the length of the cable and is relatively steady (dotted lines a), omitting numerous visible weak resonance effects. In the frequency subrange D, magnitudes of CM currents decrease with the slope of about -20 dB per decade (dotted lines), which is observed up to the frequency of the motor winding resonance frequency $f_{Rs,M}$. In between the resultant increase, ratio of the FC output CM currents achieves value of 1 for frequencies slightly lower than $f_{Rs,M}$, where the ratio of the motor feeding cable effect begins to be lower than 1, thus damping (subrange E).

Results of experimental verification of this analysis are presented in Figure 7.14 for the 100 m-long motor feeding cable. The presented CM current characteristics have been measured using a wide-band passive current probe and an EMI receiver. The effects caused by the motor feeding cable length on CM currents generated at the FC output can be seen directly on conducted emission characteristics $I_{CM,Out}(100m)$ and $I_{CM,Out}(1m)$ in which measured magnitudes of CM currents are exposed. In the bottom part of Figure 7.14), a

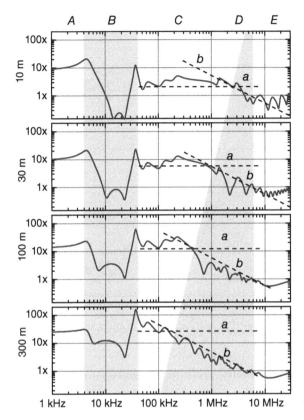

Figure 7.13 Simulation results of the motor feeding cable influence on CM currents generated at the FC output for different cable lengths 10, 30, 100, and 300 m.

relative gain factor is also presented, which can be compared with simulation results presented in Figure 7.13.

The obtained experimental results allow finding correlations between the motor feeding cable lengths and the generated CM currents at the output side of FC and define the general principle of this influence based on simulation analysis of CM impedances of the FC load (dotted lines *a* and *b*). Nevertheless, actual magnitudes of cable influence factors as a function of frequency are not reflected satisfactorily enough by the used simplified approach focused mainly on CM impedances of the FC load. To achieve better adequacy with experimental results obtained for a complete ASD, CM coupling occurring inside the FC should be also taken into consideration. Therefore, the proposed simplified circuit model based on CM impedances of the FC load can be used for general analysis of the cable length impact on conducted emission generated at

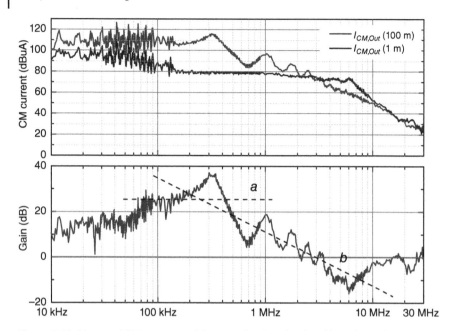

Figure 7.14 Measured CM currents at FC output for short (1 m) and long (100 m) motor feeding cables.

the output side of FC, although simulation results will still require to be verified experimentally.

Summarizing, based on investigations, effects of the motor feeding cable on CM currents generated at the output side of FC can be classified into some categories associated with frequency sub-bands defined by characteristic resonance frequencies of the motor windings $f_{Rp,M}$ and the feeding cable $f_{C,T}$ (Figure 7.15).

In the frequency range below the cable transition frequency $f_{C,T}$ (subranges *A*, *B* and *C*), constant increase of CM currents dependent on parasitic CM capacitances of the cable appears, nevertheless, in the subrange *B* as strong influence of the main parallel resonance of the motor windings $f_{Rp,M}$ occur and the level of the cable influence can be highly variable and difficult to predict by simulation. In the subrange *A*, constant increase of CM currents also occurs, but lower than that in the subrange *C*, however, from the point of view of the lower limit of the frequency range of conducted emission equal to 9 kHz, this frequency subrange is usually very narrow or is even located outside the range of conducted emission.

Above the transition frequency $f_{C,T}$ of the cable (subranges *D* and *E*), the influence of CM currents caused by the cable is strongly decreasing with frequency up to that for which the CM impedance of the motor windings achieve

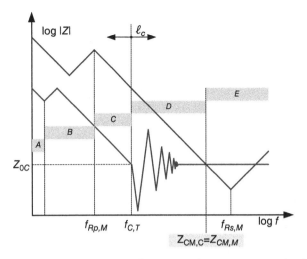

Figure 7.15 Characteristic frequency subranges associated with motor cable influence on CM currents generated at the output side of FC.

the value equal to the CM impedance of the cable, which in this frequency range is usually very close to the characteristic impedance of the cable $Z_{0,C}$. Above this frequency, which is usually slightly lower than the frequency of the main serial resonance of the motor windings $f_{Rs,M}$, cable influence decreases; thus, it is a positive phenomenon. It should be underlined, that in the subrange D, especially within the lower part, significant variations in cable influence are caused by cable resonances, which are damped according to the level of HF losses occurring in the cable.

Due to the feeding cable, the lowest CM impedances of the motor occurring in frequency range close to the main serial resonance of the motor windings $f_{Rs,M}$ are increased to the characteristic impedance of cable $Z_{0,C}$, which is usually higher than motor winding impedance for most of AC motors (Figure 6.30). Nevertheless, the decrease of CM currents within this specific frequency sub band is very much dependent on numerous resonances occurring in this frequency range, which are usually observed above 1 MHz and are very difficult to predict.

Generally, the influence of motor feeding cable on CM currents generated at the FC output can be summarized as strongly increasing within the lower part of the conducted emission frequency range, and slightly decreasing within the higher part. The cable transition frequency $f_{C,T}$ can be considered as the most meaningful parameter of the feeding cable for CM currents generated at the FC output that determines the widths of the key subranges C and D.

7.3.2 CM Currents in Motor Windings

Insertion of the motor feeding cable between the FC and the AC motor windings causes a significant increase in CM current at the FC output. One part of CM current generated at the FC output flows towards ground via CM parasitic capacitances between cable wires and grounded shield, and the second part flows between the motor windings and the grounded stator (Figure 7.1). The distribution ratio between the CM current flowing via the motor feeding cable $I_{CM.MC}$ and the motor windings $I_{CM,M}$ depends on the cable length, because the efficiency of propagation of currents and voltages via the motor feeding cable is correlated to the cable transition frequency $f_{C,T}$, above which TL effects occur in the cable. The frequency-dependent impedance mismatch between the motor feeding cable and the motor windings additionally convolve transfer characteristics of the cable.

The effect of the motor cable on CM currents injected in to the motor's windings can be considered in two ways. On one hand, it can be seen as an attenuation ratio between $I_{CM.M}$ and $I_{CM,Out}$ occurring in ASDs for the given cable length, and, on the other hand, as a ratio between the motor's CM current $I_{CM.M}$ obtained in an ASD with a very short cable and in the same ASD but with a longer cable. Distribution of CM currents between the motor feeding cable and the motor windings in the ASD with a 100 m-long motor cable is presented in Figure 7.16.

CM currents measured at the FC output and the motor's terminals are presented in the upper part of the figure, which shows that the part of CM currents generated at the FC output is lower than that in the motor winding over the entire frequency range of conducted emission. This is the effect of the fact that part of the generated CM currents is bypassed via the CM capacitance of the motor cable, and only the remaining part of generated CM currents reaches the motor's terminals. A spectral transfer characteristic of the tested motor feeding cable calculated as a ratio between $I_{CM.M}$ and $I_{CM.Out}$ is presented in the lower part of figure by the curve marked as "measured." Furthermore, in the same characteristic, curve marked as "calculated" shows results of simulation analysis carried out using simplified single-line models of the FC load.

Based on this comparison, it can be noticed that the highest levels of attenuation of CM currents in the motor cable occur between frequency ranges associated with the motor's windings main parallel resonance $f_{Rp,M}$ marked as the area A, and the cable transition frequency $f_{C,T}$ marked as the area B. For the evaluated case, it is more than 20 dB (dotted line a). Above the transition frequency of cable $f_{C,T}$, at which the motor cable exposes properties of a TL, attenuation of the cable strongly decreases and achieves much lower level, but is still positive and relatively steady (dotted line b). This frequency subrange continues up to the main serial resonance of the motor $f_{Rs,M}$, above which attenuation of the cable increases again. Nevertheless, in the frequency range

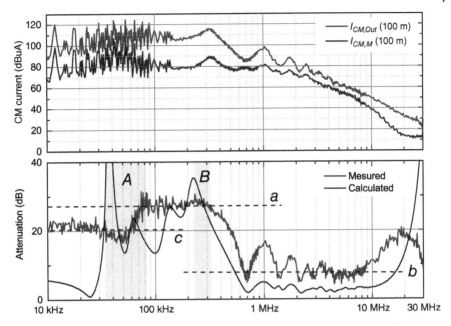

Figure 7.16 Transfer of CM currents generated at the FC output $I_{CM,Out}$ via 100 m-long feeding cable toward motor windings ($I_{CM,M}$).

between $f_{C,T}$ and $f_{Rs,M}$, the effects of signal reflection occurring in the cable result in local increases of the cable attenuation, occurring especially in the lower part of this range.

In the frequency range below $f_{Rp,M}$, attenuation of the cable for CM currents is lower, because higher CM impedances of the motor windings in this frequency range are less influenced by CM impedances of the cable, according to the formula of parallel connection of impedances. Simulation results obtained using the proposed simplified circuit modeling approach in general well reflect the characteristic frequency ranges in which the character of the cable influence changes; nevertheless, the obtained in-compliances with experimental results, especially in terms of attenuation magnitudes occurring close to resonance frequencies, are not sufficient and require further investigations. Finally, the effectiveness of filtering of CM currents initiated at the FC output by the motor feeding cable, can be considered as much higher in frequency range below the transition frequency $f_{C,T}$ of the cable, in which the cable acts as a TL. However, above the transition frequency $f_{C,T}$ of cable, elevated levels of CM current at the FC output by the motor feeding cable are transmitted effectively towards the motor windings.

An influence of motor feeding cables of different lengths on CM currents occurring in motor windings $I_{CM,M}$ can be estimated by simulation using

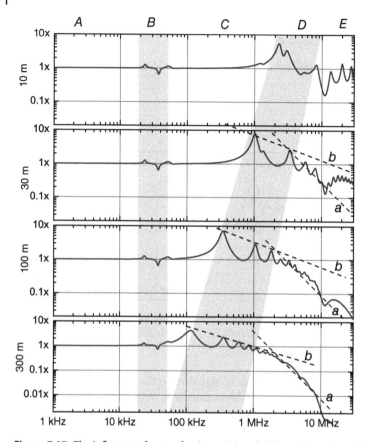

Figure 7.17 The influence of motor feeding cables of different lengths on CM currents magnitudes at motor terminals $I_{CM,M}$: simulation results.

developed simplified circuit models of motor feeding cable and windings (Figure 7.3). Comparison of simulation results obtained for the evaluated FC and motor supplied by 10, 30, 100, and 300 m-long cables is presented in Figure 7.17.

In the presented characteristics, the effects of the motor feeding cable have been determined as a ratio between CM currents of the motor occurring in the same tested ASD setup with cables of different lengths and with a very short cable. For all examined cable lengths, an increase of CM currents in motor windings have been encountered for frequencies associated with resonances resulting from cable's electrical length λ, especially $\lambda/4$, $3\lambda/4$, $5\lambda/4$ (area marked as D in Figure 7.17).

Generally, with the increase of frequency, the level of attenuation of a cable increases due to increasing HF losses in a cable; therefore, subsequent resonances of higher frequencies are more effectively damped, which is marked by dotted lines b in Figure 7.17. The attenuation effect of the motor feeding cable is also strengthened by low level of CM impedance of the attached motor windings, which in the evaluated motor occur in frequency range close to 10 MHz. Therefore, for frequencies above few megahertz, the attenuation effect of CM currents transmitted in the motor cable is strengthened by the main serial resonance of motor windings, which is indicated by dotted lines a in Figure 7.17.

The frequency range in which the elevated CM currents occur in motor windings (area D in Figure 7.17) moves towards lower frequencies for the increasing lengths of cables, which results in widening of the frequency subrange marked as E, where the influence of the cable has attenuating character, and thus is positive from the motor point of view.

The magnitudes of a gain ratio of CM currents in motor windings caused by the motor feeding cable at frequencies close to cable resonances are rather not significantly dependent on the cable length. For simulated cases with cable lengths 30, 100, and 300 m, slight decrease of maximally obtained gain can be observed, which is an effect of higher resultant attenuation of a longer cable.

Based on the results of the investigations, it can be noticed that below the cable transition frequency $f_{C,T}$ CM currents magnitudes in motor windings remain almost unchanged regardless of the length of the motor feeding cable. Only in the frequency range marked in Figure 7.17 as the area B, a small fluctuations of cable impact, correlated to frequency of the main parallel resonance of motor windings $f_{Rp,M}$, are observed.

Experimental verification of influence of the motor feeding cable length on CM currents transfer via motor windings have been done for the 100 m-long motor cable. The obtained results are presented in Figure 7.18, where magnitudes of CM currents recorded in the motor windings are compared for setups with 1 and 100 m-long feeding cables. The resultant cable gain presented in the lower characteristic was calculated as a ratio between CM currents shown in the upper characteristics.

The experimentally obtained data confirm that the motor feeding cable increases the magnitudes of motor's CM currents within frequency ranges correlated to a few lowest resonance frequencies of the cable, what was concluded from simulation analysis.

A significant attenuation of CM currents measured in frequency range above visible resonance frequencies of cable is also well correlated to results obtained by simulation using the developed circuit models (Figure 7.17). However, the limited adequacy obtained between experimental data and simulation results suggests that the proposed simplified circuit models still require further improvement to achieve better accuracy. It should also be underlined that experimental verification in such a wide frequency range and a wide range of magnitudes of measured signals is difficult and often requires the use of different measurement methods within particular frequency subranges. Frequency

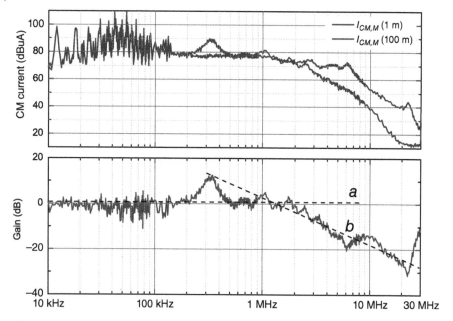

Figure 7.18 CM currents measured at motor's terminal for short (1 m) and long (100 m) feeding cables.

domain measurements in using spectrum analyzers depend on different averaging methods used in different frequency bands, and therefore uniform analysis is difficult in the whole frequency range of conducted emission. On the other hand, measurements in time domain using oscilloscope-type methods are more direct and can be preferred for high levels of signals occurring in the lower part of conducted emission band. Nevertheless, oscilloscope methods are much less immune to high levels of accompanying noises, especially in HF range.

Summarizing, motor feeding cables expose attenuation characteristics for transmitted CM currents over the entire frequency range of conducted emission. This means that CM currents injected into the motor feeding cable at the FC output are always higher than those observed in motors windings (Figure 7.16). Cable attenuation effect is not only related to cable losses but is also a result in generation of additional CM currents at the output side of the FC by cable itself, which does not reach the motor end of the cable.

Using a different evaluation approach, based on comparison of CM currents in motor windings fed via short and long cables, influence of a motor cable on magnitudes of CM currents in motor windings can be found as neutral, increasing, or decreasing, depending on frequency ranges correlated to the electrical length of the cable λ (Figure 7.18).

7.3.3   CM Currents Injected into the Power Grid

CM currents generated at the output side of FC are also transferred toward the power grid via DC bus and the input rectifier. Commonly, DC bus connection in FCs includes a very large capacitor to provide efficient filtering of DM voltage ripples generated at the DC bus by the grid-side rectifier (Figure 4.3). DC bus connection is usually equipped with DC chokes that are helpful for limiting LF harmonic current components injected into the power grid by rectifiers.

From the point of view of CM conducted emission transfer through the DC bus, the most meaningful are CM capacitances between DC buses and ground, which together with a DC bus choke, form a kind of a CM filter. CM capacitances of DC buses always exist as parasitic capacitances, but very often are intentionally increased by adding extra CM capacitors to increase the filtering efficiency. Nevertheless, because of the existence of many other parasitic couplings between internal components of FC, CM currents transfer from the FC output side to the grid side is difficult to model accurately.

The influence of motor feeding cable on grid-injected CM currents has been investigated experimentally in the evaluated ASD with a 100 m-long cable. Comparison of CM currents generated at the output side and the grid side of the ASD is presented in Figure 7.19.

On the basis of the presented characteristics, first a good correspondence between fluctuations of generated CM currents occurring in very similar frequency sub bands can be noticed, which are closely associated with the electrical length of the cable, starting from the frequency of about 300 kHz correlated with $\lambda/4$. Second, magnitudes of CM currents detected at the input side of FC are lower within almost the entire frequency range of conducted emission, starting from about 30 kHz. This means that below this frequency named as FC transition frequency $f_{CF,T}$, CM currents initiated by voltage switchings at the FC output are more effectively transmitted toward the power grid than to FC load, because the CM impedance of FC load begins to be greater than the CM impedance seen by output terminals of FC toward the power grid.

Calculated ratios between the CM current at the output side of FC $I_{CM,Out}$ and the input side of $I_{CM,In}$ are presented in the lower graph in Figure 7.19 as FC attenuation characteristics, for a 100 -mlong cable and also for a short cable. Both these characteristics are very similar to each other, which means that efficiency of transfer of CM currents from the output side to the input side of FC is not significantly influenced by CM impedance changes of the FC load caused by different lengths of motor cables.

Nevertheless, levels of CM currents occurring at the grid side of FC are finally indirectly dependent on the cable length because longer motor feeding cables increase CM currents generated at the FC output side; thus, higher levels are transferred toward the grid side, despite the fact that transfer efficiency in unchanged. Therefore, increased emission of CM currents at the FC output

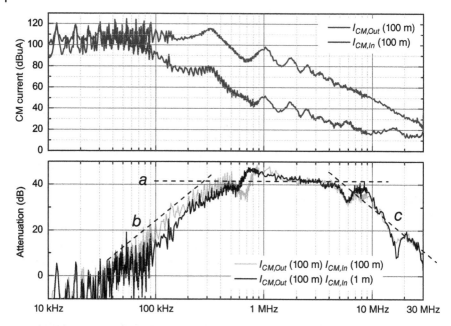

Figure 7.19 Correlation between CM currents generated at the output side of FC and injected into the power grid by the ASD with 100 m-long motor cable (upper graph) and transfer characteristics of CM currents generated at the output side of FC towards the grid side (lower graph): experimental results.

side due to longer motor feeding cables, has an effect on CM currents injected into the power grid. For the evaluated ASD with short and long motor feeding cable grid-injected CM currents have been compared in Figure 7.20.

This comparison allows noticing the frequency range approximately up to the motor windings' main resonance $f_{Rp,M}$ (frequency subrange marked as the area A), CM currents emission to the grid is significantly and steadily increased by CM parasitic capacitance of a long feeding cable in relation to a short cable. In the evaluated case, this increase was about 20 dB, indicated by the dotted line a in the lower graph in Figure 7.20.

In the higher frequency range (marked as the area B), the properties of feeding cables in HF range begin to be more meaningful. First, attenuation properties of a cable decrease as a result of amplifying effect of the cable for CM currents (dotted line b). Second, regular variations of the cable attenuation caused by reflection phenomena in a cable are clearly visible. The increase of conducted emission of almost 40 dB at frequency of the lowest resonance correlated with the length of the cable ($\lambda/4$) should be considered as a significantly negative effect of the motor feeding cable on conducted emission of the ASD. Moreover, as it was compared in the lower graph of Figure 7.20, the increase of conducted

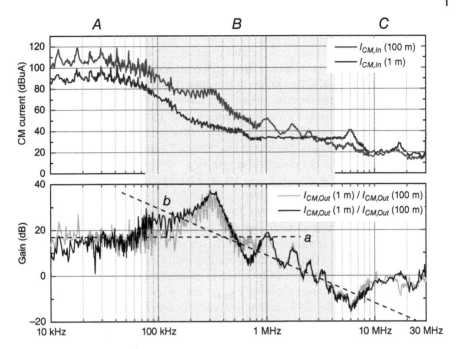

Figure 7.20 Comparison of CM currents injected into the power grid by the same ASD setup with short and long motor feeding cables (upper graph) and correlation between influence of the motor feeding cable on CM currents magnitudes at the output and grid sides of the ASD (lower graph): experimental results.

emission caused by the motor feeding cable is very similar at the output and grid sides of FC.

In the frequency range close to the main serial resonance of motor windings $f_{Rs,M}$, marked as the area C, positive effect of the motor feeding cable on conducted emission occurs at the input and output sides of FC. It is a result of an increase of the lowest CM impedance of motor windings caused by the motor feeding cable, up to the level close to the characteristic impedance of a cable $Z_{0,C}$. Effective estimation of a motor feeding cable influence above the main serial resonance of motor windings $f_{Rs,M}$ is very difficult, because of the increasing number of resonances usually occurring in this frequency range, which are not entirely reflected by the simplified circuit models. For frequencies higher than $f_{Rs,M}$, the length of motor cable is much less influential on CM currents emission and its effect can be very variant, slightly positive, but negative as well.

7.4 Summary

Motor feeding cables, as a passive and most often changeable components of ASDs, can significantly influence their EMI emission at the output side of FCs. Increased emission of CM currents at the output side of FC results in increased transfer of the CM currents towards the grid side of FC via DC bus connection.

There are two characteristic frequencies that allow identifying frequency subranges of conducted emission in which motor feeding cables impact is very significant. These frequencies are: characteristic transition frequency of a cable $f_{C,T}$ associated with the electrical length of a cable, and the main resonance frequency of motor windings $f_{Rs,M}$ for which CM impedances of the motor's windings reach the minimal values.

Between these two frequencies $f_{C,T}$ and $f_{Rs,M}$, motor feeding cable parameters are the most critical and can significantly increase the conducted emission in narrow frequency sub-bands, correlated with the cable electrical length λ. Below $f_{C,T}$, the cable influence is directly correlated to the increased CM capacitances of an FC load by wires-to-ground capacitances of a cable.

In the frequency range close to $f_{Rs,M}$, the motor feeding cable impact is positive, for example, CM current emission is decreased. Above $f_{Rs,M}$, the impact of motor cable on CM current emission is rather smaller, changes and is difficult to determine. It can be positive or negative depending upon specific distribution of parasitic couplings.

References

1 Y. Huangfu, S. Wang, L. D. Rienzo, and J. Zhu, "Radiated EMI modeling and performance analysis for PWM PMSM drive system based on field-circuit coupled FEM," *IEEE Transactions on Magnetics*, vol. 53, no. 99, pp. 1–1, 2017.

2 J. Sun, "Conducted EMI modeling and mitigation for power converters and motor drives," in *2012 ESA Workshop on Aerospace EMC*, May 2012, pp. 1–6.

3 Z. Vrankovic, G. L. Skibinski, and C. Winterhalter, "Novel double clamp methodology to reduce shielded cable radiated emissions initiated by electronic device switching," *IEEE Transactions on Industry Applications*, vol. 53, no. 1, pp. 327–339, Jan. 2017.

4 J. Luszcz, "AC motor feeding cable consequences on EMC performance of ASD," in 2013 IEEE International Symposium on *Electromagnetic Compatibility (EMC)*, 2013, pp. 248–252. Available at http://ieeexplore.ieee.org/stamp/stamp.jsp?arnumber=6670418

5 Gubia-Villabona, P. Sanchis-Gurpide, O. Alonso-Sadaba, A. Lumbreras-Azanza, and L. Marroyo-Palomo, "Simplified high-frequency model for AC drives," in *28th Annual Conference of the Industrial Electronics*

Society (IECON'02), vol. 2, IEEE, 2002, pp. 1144–1149. Available at http://ieeexplore.ieee.org/stamp/stamp.jsp?arnumber=1185434

6   Y. Ryu and K. J. Han, "Improved transmission line model of the stator winding structure of an AC motor considering high-frequency conductor and dielectric effects," in *2017 IEEE International Electric Machines and Drives Conference (IEMDC)*, May 2017, pp. 1–6.

7   G. Skibinski, R. Tallam, R. Reese, B. Buchholz, and R. Lukaszewski, "Common mode and differential mode analysis of three phase cables for PWM AC drives," in *Conference Record of the 2006 IEEE Industry Applications Conference, Forty-First IAS Annual Meeting*, vol. 2, Oct. 2006, pp. 880–888.

8   J. Sun and L. Xing, "Parameterization of three-phase electric machine models for EMI simulation," *IEEE Transactions on Power Electronics*, vol. 29, no. 1, pp. 36–41, Jan. 2014.

9   Q. Chang, Q. Ge, and B. Zhang, "A novel filter design approach for mitigating motor terminal overvoltage in long cable PWM drives," in *2014 9th IEEE Conference on Industrial Electronics and Applications*, June 2014, pp. 1798–1803.

10   Z. Liu and G. L. Skibinski, "Method to reduce overvoltage on AC motor insulation from inverters with ultra-long cable," in *2017 IEEE International Electric Machines and Drives Conference (IEMDC)*, May 2017, pp. 1–8.

11   A. Said and K. Al-Haddad, "A new approach to analyze the overvoltages due to the cable lengths and EMI on adjustable speed drive motors," in *2004 IEEE 35th Annual Power Electronics Specialists Conference (PESC'04)*, vol. 5. IEEE, 2004, pp. 3964–3970. Available at http://ieeexplore.ieee.org/xpls/abs_all.jsp?arnumber=1355176

12   R. M. Tallam, G. L. Skibinski, T. A. Shudarek, and R. A. Lukaszewski, "Integrated differential-mode and common-mode filter to mitigate the effects of long motor leads on AC drives," *IEEE Transactions on Industry Applications*, vol. 47, no. 5, pp. 2075–2083, Sept. 2011.

13   G. Skibinski, "Design methodology of a cable terminator to reduce reflected voltage on AC motors," in *Conference Record of the 1996 IEEE Industry Applications Conference, 1996, Thirty-First IAS Annual Meeting (IAS '96.)*, vol. 1, Oct. 1996, pp. 153–161.

14   A. von Jouanne, D. A. Rendusara, P. N. Enjeti, and J. W. Gray, "Filtering techniques to minimize the effect of long motor leads on PWM inverter-fed AC motor drive systems," *IEEE Transactions on Industry Applications*, vol. 32, no. 4, pp. 919–926, Jul. 1996.

15   IEC 61800-3:2017 Adjustable Speed Electrical Power Drive Systems—Part 3: EMC Requirements and Specific Test Methods, International Electrotechnical Commission Standard, 2001.

16   IEC 61000-3-6 (2008): Electromagnetic Compatibility (EMC)—Part 3-6: Limits—Assessment of Emission Limits for the Connection of Distorting Installations to MV, HV and EHV Power Systems, International Electrotechnical Commission Standard, 2008.

Appendix A

Related Standards

American National Standard Recommended Practice for Electromagnetic Compatibility Limits and Test Levels, American National Standards Institute, 2016.

IEC 61000-5-1:1996. Electromagnetic Compatibility (EMC) Part 5: Installation and Mitigation Guidelines Section 1: General Consideration, International Electrotechnical Commission.

IEC 61000-6-3. Electromagnetic Compatibility (EMC)—Part 6-3: Generic Standards—Emission Standard for Residential, Commercial and Light-Industrial Environments, International Electrotechnical Commission.

IEC 61000-6-4. Electromagnetic Compatibility (EMC)—Part 6-4: Generic Standards—Emission Standard for Industrial Environments, International Electrotechnical Commission.

IEC 61800-3:2017. Adjustable Speed Electrical Power Drive Systems—Part 3: EMC Requirements and Specific Test Methods, International Electrotechnical Commission.

IEEE Standard 519.: Recommended Practices and Requirements for Harmonic Control in Electrical Power Systems, International Electrotechnical Commission, 1992.

IEC 61000-3-6.: Electromagnetic Compatibility (EMC)—Part 3-6: Limits—Assessment of Emission Limits for the Connection of Distorting Installations to MV, HV and EHV Power Systems, International Electrotechnical Commission, 2008.

MIL-STD-461E. Requirements for the Control of Electromagnetic Interference Characteristics of Subsystems and Equipment, Department of Defense Interface Standard, August 20, 1999.

High Frequency Conducted Emission in AC Motor Drives Fed by Frequency Converters: Sources and Propagation Paths, First Edition. Jaroslaw Luszcz.
© 2018 by The Institute of Electrical and Electronic Engineers, Inc. Published 2018 by John Wiley & Sons, Inc.

Appendix B

Further Readings

G. Benysek, *Improvement in the Quality of Delivery of Electrical Energy Using Power Electronics Systems*. London: Springer, 2007.

M. Bollen and F. Hassan, *Integration of Distributed Generation in the Power System*. Wiley-IEEE Press, 2011.

S. Caniggia and F. Maradei, *Signal Integrity and Radiated Emission of High-Speed Digital Systems*. John Wiley & Sons, Ltd, 2008.

F. Costa, C. Gautier, E. Labour^', B. Revol, F. Costa, C. Gautier, E. Labour^', and B. Revol, *Electromagnetic Compatibility in Power Electronics*. John Wiley & Sons, Inc., 2013.

J. Guzinski, H. Abu-Rub, and P. Strankowski, *Variable Speed AC Drives with Inverter Output Filters*, IEEE Press Series on Power Engineering. John Wiley & Sons, Inc., 2015.

E. B. Joffe and K.-S. Lock, *Grounds for Grounding: A Circuit to System Handbook*. Wiley-IEEE Press, 2011.

F. Labrique and J.-P. Louis, *Power Electronic Converters*. John Wiley & Sons, Inc., 2013.

E.-P. Li, *Electrical Modeling and Design for 3D System Integration: 3D Integrated Circuits and Packaging, Signal Integrity, Power Integrity and EMC*. Wiley-IEEE Press, 2012.

H. W. Ott, *Electromagnetic Compatibility Engineering*. John Wiley & Sons, Inc., 2009.

High Frequency Conducted Emission in AC Motor Drives Fed by Frequency Converters: Sources and Propagation Paths, First Edition. Jaroslaw Luszcz.
© 2018 by The Institute of Electrical and Electronic Engineers, Inc. Published 2018 by John Wiley & Sons, Inc.

C. R. Paul, *Introduction to Electromagnetic Compatibility.* John Wiley & Sons, Inc., 2006.

R. Perez, *Handbook of Electromagnetic Compatibility.* Academic Press, 1995.

R. Smolenski, *Conducted Electromagnetic Interference (EMI) in Smart Grids.* Springer, 2012.

R. Strzelecki and G. Benysek, *Power Electronics in Smart Electrical Energy Networks.* London: Springer, 2008.

F. Zare, Ed., *Electromagnetic Interference Issues in Power Electronics and Power Systems.* Bentham Science Publishers Ltd, 2011.

INDEX

High Frequency Conducted Emission in AC Motor Drives Fed by Frequency Converters: Sources and Propagation Paths, First Edition. Jaroslaw Luszcz.
© 2018 by The Institute of Electrical and Electronic Engineers, Inc. Published 2018 by John Wiley & Sons, Inc.